DYNAMICS OF RESIDENTIAL WATER DEMAND AND SUPPLY IN INDIA

DYNAMICS OF RESIDENTIAL WATER DEMAND AND SUPPLY IN INDIA

A Case Study of Chennai City

Dr. A Munian

GYAN

The Publication was financially supported by the ICSSR and the responsibility for the facts stated opinions expressed or conclusions reached, is entirely that of the author and that the ICSSR accepts no responsibility for them.
Consultant Professor Amitabh Kundu, JNU, New Delhi

Dynamics of Residential Water Demand and Supply in India : A Case Study of Chennai City

ISBN: 978-81-212-1062-1

Published in 2010 in India by
Gyan Publishing House
23, Main Ansari Road, Darya Ganj
New Delhi-110002
Phones : 23282060, 23261060
Fax : 011-23285914
E-mail: books@gyanbooks.com
Website: www.gyanbooks.com

Laser Type Setting by: Rudra Computer Graphics, Delhi
Printed at : Chawla Offset Printers, Delhi

Contents

List of Tables

List of Figures

Preface

The need for supplying potable water to people through organised water supply stems from the fact that water as an essential requirement inextricably linked to the state of human health. Water is identified as an indispensable resource for the maintenance of life. Though water endowment is more or less constant over the years, some parts of the globe face acute water scarcity owing to rapid population growth, unplanned urbanisation, mindless industrialisation and water pollution. As a result of which there is acute scarcity of water in many cities. Both natural and manmade actions are responsible for water scarcity.

Water reliability considers water as a commodity and ensures households to have full or firm control over available water. Hence, reliability depends on its physical location and timely availability. Water usage related to entitlement rights of households to meet their basic needs viz., drinking, cooking, hygiene, sanitation and the supplementary needs. Unreliable water supply requires households to adopt various coping or alternative strategies to compensate unreliability in water supply. Coping with periodic or seasonal scarcity is costly due to the requirements of investments in water storage equipment, water-treating gadgets, and the opportunity cost of time spent on fetching water.

There are still millions of people in the world who do not have access to safe drinking water. The poor are the worst affected in this regard. For poor, access to potable water may involve trekking long distances, long waits, and hiring trips back home with jerry-cans. Water chores take up a substantial part of women's and girls' time. This is because of the fact that the responsibility of bringing water from long distances traditionally falls on women and girls. According to the "Human Development Report -2006" of the United Nations Development Programme, about 1.1 billion people in the developing world do not have access to even the minimal quantity of clean water.

Chennai being the fourth largest metropolitan city in India with over 4.7 million inhabitants was supplied with 300 mld of potable water per day in 2004. This is principally drawn from three reservoirs located at the periphery of Chennai City and fluctuating supply is adjusted with the water extracted from local aquifer and the remainder being piped from a long distance source called Veeranam Lake. The bulk of the water is being used for domestic purpose and the rest being taken for commercial, service and industrial purposes.

Chennai City, the capital of Tamil Nadu State has been facing critical depletion of water resources to an increasing growth of its population. The reckless extraction of water from depleted aquifers, scarcity of surface water due to insufficient rains, water contamination and a long list of service management deficiencies lead to a losing fight for the metro water authorities to provide the basic service of water supply to the entire population in Chennai City. On the one hand there is an ever-increasing demand for water at the one hand and on the other the quantity of available water dwindles as well as its quality declines. Demand is also restricted because of water shortage; piped water supply does not reach significant portions of the city people on a regular basis. This situation leads to unreliability of water supply in Chennai City. The average per capita availability of water around 70 lpcd during the normal years is the lowest for any metropolitan city of India. In the present context, it is critical to conduct a comprehensive study of the unreliability of water supply, compensatory strategies used by the households and the cost of unreliability. The estimation of unreliability of water supply will help to understand the need for more investment in order to reduce the social cost of water supply in Chennai City.

It is hoped that the book will be very useful for administrators, policy makers, urban planners, researchers and others involved in the field of urban water supply and development.

Dr. A. Munian

Acknowledgements

I would like to express my gratitude to those individuals for their wonderful support and assistance during the course of completing this book. Dr. B.P. Chandramohan, M.A., M.Phil., Ph.D., Associate Professor has been a strong and supportive supervisor to me throughout my studies at Presidency College, Chennai-05 giving me great freedom to pursue independent work. More importantly, he demonstrated his faith in my ability and encouraged me to rise to the occasion. He has always taken time to introduce me to the people within the discipline, kept me focussed, carefully listened to my problems, and has been a flexible but a strong advocate to me.

I have been especially fortunate to work with such a great following individuals and institutions also deserve special mention for their contributions to this book and their support, which is gratefully acknowledged.

My special thanks to the faculty members, Department of Economics, Presidency College, Chennai – 5 for having approved my research work. I express my sincere thanks to Managing Director, Chennai Metropolitan Water Supply and Sewerage Board (CMWSSB) for having given permission to collect data and encouraged me in completing my book.

I kindly acknowledge the invaluable financial assistance provided by the Indian Council Social Science Research (ICSSR), Ministry of Human Resource Development, Government of India, New Delhi for the publication of my book.

I express my sincere thanks to ICSSR consultant and Professor Amitabh Kundu, Jawaharlal Nehru University, New Delhi who evaluated my research work and recommended to ICSSR for the publication of book.

A special thank goes to my parents, my brothers and sisters, my wife and for their unconditional support, patience and love which was always there for me, and also My daughter M.Anitha for her amazing adventures, which kept me inspired every day. The author would also like to thank Gyan Publishing House, New Delhi, India for having given permission to publish my book.

Finally, in recollecting the memories of past few years, I am delighted to see the sheer scale of excitement, frustration and rich rewards that this work has brought me, and hopeful that I would have an opportunity to share this wealth of experience and knowledge of my work with others.

Dr. A Munian

1

Introduction

Water is identified as an indispensable resource for the maintenance of life. Though water endowment is more or less constant over the years, some parts of the globe face acute water scarcity owing to rapid population growth, unplanned urbanisation, mindless industrialisation and water pollution. As a result of which there is acute scarcity of water in many cities. Both natural and manmade actions are responsible for water scarcity. The World Commission for Water (2000) estimates that more than 1 billion people in developing countries do not have access to clean water. Available evidences show that the per capita water availability has declined by 40 per cent between 1950 and 1990 (Ayibotele, 1992). In addition to population growth, changing demographic characteristics often affect the management of urban water resources. Water supply crisis forced many urban authorities to over-exploit the fragile source such as aquifers as a short-term remedy, which ultimately resulted in falling ground water, levels in many cities. Deterioration of water supply and its effect on sanitation leads to a progressive decline in urban living conditions.

Population growth, increase in GNP and progressive industrialisation combined together to create a demand for water to the level of crisis in the urban areas of developing countries, which will continue to increase substantially in the coming decades. The rate of population growth in urban areas far exceeds that of the rural areas. Lack of clean water is the most serious health problem in the low-income settlements. The direct effect of poor water supply is the prevalence of diseases, while the indirect effects include lost earnings and lost educational opportunities for young people.

According to the Central Public Health Engineering Organisation, (2000) 88 per cent of the urban population in India has access to potable water supply. But this supply is highly erratic and unreliable. Transmission and distribution networks are old and poorly maintained and they result in poor quality of water. Consequently, physical losses are very high, ranging from 25 to over 50 per cent. Low pressures and intermittent supplies allow back siphoning, which results in contamination of water in the distribution network. All this is even without considering the quality of water supplied. Inadequate and unreliable water supply service in metropolitan cities of developing countries affects the life of millions of urban dwellers. The huge volume of water and extensive infrastructural requirements often create problems in fulfilling adequate demand for water.

The water supply sector is confronted with a number of problems and constraints. According to the views of WHO, an urban dweller has access to potable water if he is served by a piped network or stand post 200 metres away from his dwelling. Even though the accessibility has improved over the years, still a majority of the urban inhabitants are still deprived of access to safe water. Water security encompasses issues of water availability; accessibility usage and water quality.

Water supply unreliability is the absence of water security although availability alone does not ensure water security at a given point in time.

Water reliability considers water as a commodity and ensures households to have full or firm control over available water. Hence, reliability depends on its physical location and timely availability. Water usage related to entitlement rights of households to meet their basic needs viz., drinking, cooking, hygiene, sanitation and the supplementary needs like washing, watering etc. Unreliable water supply requires households to adopt various coping or alternative strategies such as use of multiple sources of water and commuting long distances to access water. Coping with periodic or seasonal scarcity is costly due to the requirements of investments in water storage equipment, water-treating gadgets, private water purchase opportunity cost of time spent on fetching water.

If the water supply is unreliable, households will not remain passive but they adopt or develop compensatory strategies like buying private water, digging tube wells, purchasing containers or making provisions for storage of municipal water. Nevertheless this behaviour is conditioned by socio-economic characteristics and availability of alternative water sources. Household income may be an important factor determining the service level conditions. Those households that suffer deficient service conditions may be mostly living in slums.

Urban poor in metropolitan cities pay relatively heavy prices for water besides incurring a high opportunity cost. Inadequate coverage and decline in urban infrastructure hit the poor the hardest as wealthier households tend to have access to alternative water supplies or they can afford tube wells of their own if the supply is unreliable. Women and children in low-income settlements spend considerable time and energy in collecting the minimum required quantities of water from sources at long distances from their dwelling. In most of the cities availability of drinking water is becoming progressively scarcer and more uncertain. The search for drinking water from distant places led to unequal and uncertain access to water resulting in conflicts. Most of the cities suffer from a water crisis caused by inappropriate management of the water infrastructure and distribution problems.

The World Bank estimates that the urban population of India which is around 285 million inhabitants in 1991, accounted for 50 per cent of the country's Gross Domestic Product (GDP). Hence, growth of GDP and growth of tax base depend on improving urban infrastructure (Economic Survey 2003-2004). The urbanisation process is not complete, because urban growth clearly accounts as one of the reasons for inadequate and poorly maintained infrastructure with high level of physical lòsses, around 30 per cent for most cities. Requirements of raw water resources increase all the more because of lack of focus on streamlining the water supply networks. Consequently, water is not available to everyone, the distribution is intermittent, number of hours is limited, pressure is low and standard water quality is not ensured. This means there is a need for new and more investment according to the estimates of Indian Planners (World Bank 1996). The annual investments including operation and maintenance required for basic infrastructure for the 1996-2001 periods would be around USD13 billion. But there is an enormous gap between the expenditure requirements and the availability of resource. It is estimated in the same study that at USD 3.5 billion a year it is just enough to cover the annual costs of operation and maintenance.

The urban local bodies have often grossly subsidised water supplies in order to achieve social and health benefits of low income households who form a majority of the population. In this process governments could not stop the benefits of subsidised water to the wealthier households. Nevertheless the richer households receive relatively reliable water while the poorer households benefit in a less than proportionate manner. The drain of resources in the name of subsidy hampers the ability to expand and improve the services provided to the poor.

Review of Literature

The relevant empirical and theoretical literature for the present study is reviewed under three broad heads: viz., Residential water demand, water supply unreliability and willingness to pay for maintaining the water supply and for its improvement.

Subrata Chattopadhyay (2001) in his study analysed that there are visible benefits to health when people have access to 200 lpcd (litre per capita day) for drinking, basic personal hygiene and cooking. He stressed that the minimum requirement of 50 lpcd must be ensured in order to avoid serious health damage. World Bank in one of its studies (1996) of 27 Asian cities with a population of more than 1 hundred thousand, it revealed that Chennai is the worst city in Asia in terms of hours of water availability per day. P.B. Anand (2001) in his study of Chennai City gave a new interpretation to water scarcity. He opined that it is a problem of there being enough water but not enough money or technology or human resources to bring that water to the people.

Ratna Reddy (1996) conducted a study on the demand for water in Rajasthan. The study indicated that the per capita water use in actual situation is much less than the prescribed norms of WHO. Slums dwellers and other low-income groups depended more on stand posts or hand pumps where the supply levels were low, unreliable and very poor. The study also evidenced that as the size of the household increases, the per capita demand would tend to decline. Jacob John (1997) studied water supply planning, water demand and pricing and economics of public water supply systems of Kerala. In his study he found that a large number of piped water supply and hand pumped schemes remain unserviceable, inoperative and redundant. The important reasons, for such a redundancy were defective design, faulty installations, poor maintenance and lack of participation.

One of the water studies published in Urban Age (1999) described the nature and causes of leakage of water when the water supply is connected to the households. Leaks often occur at joints or at house service connections due to corrosion and other structural problems. To compare in Mexico City loss of water due to leakage was estimated as12 cubic metres per second. Riordon, Courtney (1973) developed a more general model of investment pricing decisions in relation to the particular problems of choosing the timing and sizes of additions to capacity in urban water supply system. He framed a model on the basis of empirical data by using hypothetical cost and demand curves of water supply under varying assumptions with regard to rate of growth of demand, the level of the discount rate and the length of the planning horizon.

Armstrong, R.D and C.E. Willis (1977) developed models for water planning that simultaneously consider investment and allocation decisions. The World Health Organisation (WHO) claimed that a large majority of people in developing countries grossly suffers from poor access to safe water and inadequate means when compared to the industrialised countries. Geoffrey.D. Datoelko (2002) compared the size of the population with the supply of fresh water because both are widely recognised as the two important correlates deciding the demand and supply conditions of water. Sharma, Dev (2000) pointed out that only a small proportion of the world population, mostly in developed countries face low water stress and good coping capabilities. The high stress and low coping capabilities are the characteristics of developing countries with the worse hit countries being those in the arid and semi-arid zones of the world.

Vande, Hock, Wim (2001) highlighted the environmental problems due to water scarcity. Lack of sufficient access to water for drinking and bathing will increase the spread of many water-borne and communicable diseases, especially, diarrhoeal and skin diseases. Foster, Stephen (1999) studied groundwater pollution. His study pointed out that most of the cities rely on ground water for their domestic water supply. According to the World Resources Institute Studies (1996) almost all coastal cities face seawater or salination intrusion because of aquifer depletions.

According to Global Report on Human Settlements (2001) the burden of water collection for un-piped households seems to have increased during the last thirty years. The study looked at the use of water consumption, hygiene and amenities in domestic life. It also examined the cost of water in monetary terms as well as amount of energy and time spent. Hardoy.J.E, Mitlin.D and Satterth Waite,D (1990) analysed water problems subject to urban population growth, water supply, urban pollution and waste problems. They also pointed out that already many cities are in unhealthy environments with open drains, poor refuse collection and contaminated water supplies. Chris Park, (1997) indicated that although global demand for water continues to rise, the development of water resources in some countries has been restricted by regional water shortages and environmental damage associated with existing water schemes.

Lee and Anas (1989) were the pioneers in the study of unreliability. The study revealed the importance of undertaking investment in infrastructure and the inequalities in the provision of infrastructure between small and large companies. The study recommended the large companies to have its own water supply system and to sell the surplus water to smaller companies.

The World Bank (1996) conducted four independent country specific studies on unreliability. These studies concluded that households with in-house connections made long term investments to solve the problems of water quality and water supply unreliability.

M.H. Zerah, (2000) conducted a study on water supply unreliability problems in Delhi. She has narrated the overall achievements and shortcomings of water supply in Delhi. She concluded that only 5 per cent of the households receive a supply that can be compared to the standards of developed countries. She claimed that in the near future water would certainly become insufficient.

Research Problem

Chennai being the fourth largest metropolitan city in India with over 4.7 million inhabitants was supplied with 300 mld of potable water per day in 2004. This is principally drawn from three reservoirs located at the periphery of Chennai City and fluctuating supply is adjusted with the water extracted from local aquifer and the remainder being piped from a long distance source called Veeranam Lake. The bulk of the water is being used for domestic purpose and the rest being taken for commercial, service and industrial purposes.

Chennai City, the capital of Tamil Nadu State has been facing critical depletion of water resources to an increasing growth of its population. The reckless extraction of water from depleted aquifers, scarcity of surface water due to insufficient rains, water contamination and a long list of service management deficiencies lead to a losing fight for the metro water authorities to provide the basic service of water supply to the entire population in Chennai City. On the one hand there is an ever-increasing demand for water and on the other the quantity of available water dwindles and quality declines. Demand is also restricted because of water shortage; piped water supply does not reach significant portions of the city people on a regular basis. This situation leads to unreliability of water supply in Chennai City. The average per capita availability of water around 70 lpcd during the normal years is the lowest for any metropolitan city of India. According to The Hindu, July, 7th (2000) piped water supply was only 59 lpcd to the households of Chennai City in 2000. Chennai City, without a perennial source of fresh water, supplies 78 lpcd during normal years of rainfall, however, it supplies only 32 lpcd during drought years. Several times the reservoirs of Chennai City store 500 million cubic feet (Mcft) but drying up of the reservoirs is also not uncommon almost every alternative years. Variation in the availability of water across different sections of people is also considerable.

The Chennai Metropolitan Water Supply and Sewerage Board (CMWSSB) is the monopoly provider and operator of potable water for the entire categories of households with and without in-house water connection in Chennai City. Households without in-house connection fetch water from public source. The households without in-house connection are exempt from water tariff. Even the existing water tariff of the households with in-house water connection is highly subsidised.

Chennai City is one of the most water-starved cities of the world. The average lpcd (litres per capita consumption per day) of Chennai City is the lowest of all the important metropolitan cities of India. Therefore, a study of water supply unreliability in Chennai City is ideal. The consequences of water supply unreliability have been faced by households in all the classes of economic status, though they vary in degrees.

The deficit of resource has limited the urban local government both to improve the current service in those areas and to invest the necessary resources to guarantee sustainable supply. The issue of sustainability is far from being one of environmental rhetoric. The supply of water to Chennai City is largely based on the surface water availability of Chennai reservoirs. The water storage in the reservoirs depends upon the availability of rainfall. Single failure of monsoon will jeopardise the water supply situation. The Metro Water Board always undertakes ground water exploitation as the alternative strategy to meet out the water crisis. The acute water demand has caused a considerable reduction in the ground water resources available in the region. Local aquifers have been experiencing critical over exploitation. The water crisis is currently somehow managed by the distant pipeline water scheme of New Veeranam. With the growth of population, water scarcity will push the government to improve efficiency by implementing programmes to repair pipe leaks and demand side management. Water meters have been fixed to an insignificant proportion of the households in Chennai City, which lead to some negative results in demand side water management in terms of water consumption and improving collection performance. Governmental efforts to improve water supply address only the supply side of the problem. Sustainability issues require information concerning the demand side of the water market.

The frequent unreliable supply of water in terms of water shortages and low water pressure has made storage tanks a common feature of majority of households across Chennai City. The water tanks are installed on rooftops and are used to store water when pressure in the system is inadequate. Throughout different zones of Chennai City the population has adopted a range of different averting behaviours in the face of uncertain water quality standards. Similarly, boiling, filtering or disinfecting tap water, buying bottled or consuming soft drinks are the common practices.

When ground water becomes contaminated and other sources of drinking water are not available, households drill new wells or buy water for their needs. Since, a larger proportion of the households do not have in-house water connection, they depend on public water tap services. Hundreds of households rely on single tap or public hand pumps, which results in long queues to get two or three pots *(Kodams)* of water. Many times even the households with in-house water connection have inadequate or no supply. Therefore even a households connected to a metropolitan water supply network could face the problem of water supply unreliability. In addition to the problem of scarcity of supply there are certain areas within the city with the problem of discrepancies in service levels resulting in unequal distribution of water supply. It can be observed in Chennai City that water vending, retailing and tanker lorry water supply are the most lucrative business especially during the summer months. The consequences of insufficient water on the poorest households have been very pathetic. The water problem is not only of insufficient quantity but also of poor quality, which have serious consequences on their health.

Scope of the Study

Many researchers and government departments have undertaken studies in the form of projects and reports about water supply in Chennai City. Some of the studies have been conducted in water scarcity and some others on unequal distribution of water supply in Chennai City, (Kumaraswamy.T (1990); Mohanakrishnan.A (1996); Viswanathan. S, (2004). Only a few studies based on the contingent valuation methods have been undertaken around the world to evaluate the users demand for water and to estimate their willingness to pay for improved services, Vaidyanathan A and Saravanan, J. (1997); Anand P. B.; (2001). No specific study has conducted in Chennai City dealing with water supply unreliability.

In the present context, it is critical to conduct a comprehensive study of the unreliability of water supply, compensatory strategies used by the households and the cost of unreliability. The estimation of unreliability of water supply will help to understand the need for more investment in order to reduce the social cost of water supply in Chennai City.

Objectives

The overall objective of the study is to determine the residential water demand and to assess the water supply unreliability results the households to resort to different compensatory strategies. The specific objectives of the study are as follows.

1. To determine the total composite water demand of households with in-house water connections and without in-house water connections in Chennai City.

2. To identify the factors those determine household water demand for in-house connections and without in-house connections of metro water in Chennai City.

3. To estimate the average cost of unreliable water supply and to understand its relationship with various factors associated with household water consumption.

4. To study and compare the alternative strategies of households which seek to resolve the difficulties of unreliable water supply in Chennai City.

5. To estimate the households Willingness to Pay (WTP) for the maintenance of existing and improved water supply in Chennai City.

Hypothesis

1. Households with in-house water connection consume more quantity of metro water than that of the households without in-house water connection.

2. Household income influences the volume of metro water consumption.

3. Household income and area of the house are inversely related to average household annual unreliability cost.

4. Higher income households have efficient compensatory strategies.

5. Household WTP for new improved hypothetical water supply is more than the current water supply.

6. More percentage share of households without in-house water connection WTP for improved water supply as compared to households with in-house water connection.

METHODOLOGY

Sources of Data

The objectives of the study stated above have been verified with the use of both primary and secondary data. The necessary secondary data have been obtained from various published works such as books, reports, journal articles, magazines, periodicals and electronic web materials. In addition to this the researcher has made use of the data and other information pertaining to water supply, income and expenditure of water board, population growth and others from CMWSSB, Corporation of Chennai, and Chennai Metropolitan Development Authority (CMDA). The researcher has also utilized the library facilities of Institute of Financial Management and Research (IFMR), Madras School of Economics (MSE), Madras Institute of Development of Studies (MIDS), Tamil Nadu Pollution Control Board (TNPCB) and Murugappa Research Institute. Besides using the materials of various government departments and research institutions, experts in the field of water studies have also been consulted.

The secondary data have been used for focusing more on multifaceted water supply conditions in Chennai City. The objectives of the study have been verified with the help of primary data collected through schedules administered to the respondents. This process has been begun with the selection of randomly selected sample households in Chennai City.

A randomly representative sample of households in Chennai City have been selected and the schedule based interviews have been conducted to generate required information with regard to water demand, water supply unreliability and the alternative strategies adopted by the sample households to overcome the water supply unreliability.

Sample Design

The development of a good sampling frame is an important approach to identify the representative samples. The sampling method adopted is one of multi-stage sampling. Chennai City has ten zones and all the zones have been selected on Census basis in order to give equal representation to all areas of Chennai City. Each zone of Chennai City has a number of Corporation divisions as earmarked by the Chennai Corporation. At the first stage a corporation division in each zone has been taken as the sampling unit. On the probability-proportional size basis the sample corporation divisions have been selected from each zone. It implies that more divisions will be selected as samples in a zone if it has more number of divisions. At the second stage a sample street has been selected as the sampling unit. Thus on probability proportional basis the numbers of sample streets have been obtained from each sample division. At the third stage a household from the sample street has been selected as the sampling unit. The number of sample households from each sample street has been selected on probability proportional basis. It means that a sample street with more number of households will get more sample households. The selection of households from each sample division has been stratified on the basis of slum and non-slum areas.

The sampling procedure resulted in the selection of 450 households from 231 streets and the 231 streets from 25 corporation divisions in 10 corporation zones. This scientific approach has led to the selection of the samples from Chennai City comprising both slum and non-slum areas. Table 1.1 shows

Table 1.1. Selection of Samples of Chennai City

Zone No.	Zone Name	Division No.	Selected Division	No. of Streets	Sample Streets	Sample Household	Schedule No.
I	TONDIARIPET	1-13	3	70	12	23	1-23
			10	62	11	22	24-45
II	BASIN BRIDGE	14-31	14	73	8	15	46-60
			22	75	8	16	61-76
			31	69	7	14	77-90
III	PULIANTHOPE	32-49	34	100	11	22	91-112
			36	44	5	10	113-122
			40	65	7	13	123-135
IV	AYNAVARAM	50-63	51	93	6	12	136-147
			62	253	17	33	148-180
V	KILPAUK	64-78	68	73	10	20	181-200
			75	95	13	25	201-225

(Contd.)

(Contd. Table 1.1)

Zone No.	*Zone Name*	*Division No.*	*Selected Division*	*No. of Streets*	*Sample Streets*	*Sample Household*	*Schedule No.*
			79	38	8	15	226-240
VI	ICEHOUSE	79-96	88	17	9	17	241-257
			91	34	7	13	258-270
			97	24	4	7	271-277
VII	NUNGAMBAKKM	97-113	104	71	11	22	278-299
			113	50	8	16	300-315
			117	75	7	13	316-328
VIII	KODAMBAKKAM	114-129	120	64	6	12	329-340
			128	116	10	20	341-360
IX	SAIDAPET	130-141	131	121	12	23	361-392
			141	117	11	22	393-405
			153	232	14	27	406-413
X	ADYAR	142-155	155	149	9	18	414-450
	Total	155	25	2180	231	450	450

the procedure of the selection of samples from the sample areas of Chennai City.

Developing the Questionnaire

After making a preliminary analysis of the practices of supplying water to different residential areas of Chennai City, a questionnaire has been developed consisting of different sections, viz., (1) Identification Particulars (2) Background Information, (3) Household Particulars (4) Household Demand for Water (5) Water Facilities (6) Water Supply and Unreliability (7) Compensatory Strategies (8) Perception about Water Quality (9) Contingent Valuation (CV) questions on WTP and attitude towards administration of water supply services. While designing the questionnaire attempts have been made to minimize strategic, hypothetical and compliance biases, which might have arisen from contingent valuation survey. In order to avoid starting point bias price for the bidding game in administering the CV survey five different clusters of prices with different quality of services have been classified.

Model Specification

To compare the determinants of demand for metro water and total water consumption two separate Ordinary Least Square (OLS) step-wise multiple regression models have been used. Per capita household consumption of metro water per day has been taken as the criterion variable for the first model with 9 predictor variables. For the second model per capita household total consumption water per day as the criterion variable along with 6 predictor variables has been taken. The estimation has been done through the SPSS software.

In order to make the difference in the preference between the existing and the anticipated new water supply and to determine the factors influencing household preferences, a discrete economic model (Logit Multi-Regression) has been used. In this model, water services are expressed as a function of several attributes. This model describes the probability that particular households will choose to use a new water source. It is assumed that a household chooses between two sources. The first describes the utility gained from using the new service and the second is the utility derived from the current one. This is estimated by using the Logit Regression model in LIMDEP software.

Limitations of the Study

The study is limited to estimate the residential water demand, water supply unreliability and compensatory strategies used by the households in Chennai City. It does not attempt to objectively estimate the quality of water supplied by the Chennai Metro Water Board.

The present study is restricted to the domestic consumption of water. Water supply to industrial, commercial, institutional purposes is not estimated due to the limitations of time and resources.

Chapter Outline

First chapter reviews the basic problem of residential water demand, unreliable water supply and compensatory strategies adopted by urban households'. It highlights the objective of the study, hypothesis, methodology and limitations along with the details of chapter outline.

Second Chapter is concerned with the challenges for Urban Water Supply in India. It includes urban water management, Urban Water distribution and water driling.

The third chapter focuses on the historical background and development of drinking water supply in Chennai City. It includes the origin, historical development, and growth of population, urbanization, and urban growth. It also evaluates of the Chennai Metropolitan Water Supply and Sewerage Board (CMWSSB).

Fourth chapter focuses on the water resources management. It includes both the water supply and demand management. On the supply side management, this chapter discusses the existing water supply sources and water production such as surface and ground water. It also concentrates on water requirements, water supplied to domestic use, water crisis, and to analyse alternative water sources to fill the gap between demand and supply.

Fifth chapter is concerned with the socio-economic characteristics of the sample households in Chennai City inclusive of their housing characteristics. The topics covered are the demographic and general characteristics of the sample households and the aggregate of their membership, economic condition of households, housing conditions with respect to various facilities, education and cultural interests.

The determinants of the demand for metro water and the total water consumed by the households have been discussed in the Sixth chapter. It also estimates the per capita per day consumption level of metro water and total water consumed by different sections of people in Chennai City.

The Seventh chapter deals with water supply unreliability, compensatory strategies used by the households, unreliability cost of water supply and willingness to pay for the current and the hypothetical package of improved supply. The final chapter ends with giving the summary of the findings and suggestions.

2

Challenges for Urban Water Supply in India

Growth of Urbanization

Urban population of developed and developing countries is increasing at a constant growth rate. India is no exception to the prevailing trend. Urban development in India is presently going through a very dynamic stage. In the last four decades, India's urban population increased from 10.84 percent in 1901 to 25.72 percent in 1991, and 27.70 percent in 2001. The total population and urban population of India is explained Table 2.1.

The Urban population increased almost ten times between 1901 and 2001 and the number of urban settlements doubled in this period. According to (GOI 2001) in India population reached 1029.02 million of which about 742 million live in rural areas and 285 million in urban areas. The decadal growth of urban population during 1911 was 0.35 percent and it increased to 31.39 percent in 2001.

Structural adjustments, micro economic reforms and globalization of Indian economy are likely to accelerate the pace

Table 2.1: Urbanization in India

Census Year	*Total Population (Million)*	*Urban Population (Million)*	*Number of Towns*	*Urban Population as % of Total Population*	*Decimal Growth of Rate of UPC %*	*Annual Exponential Growth Rate*
1901	238.40	25.85	1827	10.84	-	-
1911	252.09	25.94	1815	10.29	0.35	0.03
1921	252.32	28.09	1949	11.18	8.27	0.79
1931	278.98	33.46	2072	11.99	19.12	1.75
1941	318.66	44.15	2250	13.86	31.97	2.77
1951	361.09	62.44	2843	17.29	41.42	3.47
1961	439.24	78.93	2355	17.17	26.41	2.34
1971	548.16	109.11	2591	19.91	38.23	3.21
1981	683.33	159.46	3378	23.34	46.14	3.83
1991	844.32	217.18	3368	25.72	36.19	3.09
2001	1029.02	285.35	5161	27.70	31.39	2.71

Source: Census of India 1991 and 2001.

of urbanization. Cities are likely to become fulcrum around which trade, commerce and industrial activities would gravitate this accelerating the process of migration and number of migrants to urban centres. The growth of urban has undoubtedly contributed to economic growth by providing economics of scale. The share of these urban areas was only 29 percent in 1950-51, 47 percent in 1980-81 and it increased to 60 percent in 2001. This could be explained in Table 2.2.

Table 2.2: Contribution of Urban areas to Economic Growth in India

Year	*Percentage of Urban to total Population*	*Estimated contribution to National Income (Percentage)*
1951	17.3	29
1981	23.3	47
1991	25.7	55
2001	30.5	60

Source: Urban India, Ministry of Urban affairs and employment, 2001.

Urban Infrastructure India

In the beginning of the 21st Century, more than half of the total world population is now living in cities and towns. India faces an unprecedented challenge in throwing up sound institutions to cope with the rapid growth of cities. Thirty crores of India's population is in cities and it is expected that by 2025 it will be fifty crores in 2015. Because of the rapid growth of urban population, cities are facing the economic, socio and environment problems such as, poverty, health hazards etc. Indian cities are rate among the lowest quality of life in the world. Urban infrastructure consists of drinking water, sanitation, sewage systems, electricity and gas distribution, urban transport, primary health services, and environmental regulations. According to 74th Constitution Amendment, many of these services are fully empowered to the local governments. The urban areas have two side effects, output and poverty. The urban areas are contributing major share to gross domestic product growth. Nearly 50 percent of India's GDP is presently estimated to come from the urban areas (Economic survey 2003-2004). Indian Economic Growth largely depends on improving public services in the cities. On the other hand, the growth of urbanization creates poverty in the cities. The poverty eradication programmes are linked to efficient systems of public services delivery. Many studies suggest that without reforms in the institutional framework for urban infrastructure, centre or state level government funds directed into urban sector will not have the economic and social returns.

Problems In Urban Services

Public services are fragmented between different corporation, agencies and local government bodies of each state with conflicting lines of accountability. These agencies are providing municipal service to the city people with limited resources, limited accountability no incentives for innovation in delivery of public services. But there is a nexus between political and administrative systems and poor people are not able to obtain municipal services.

Fiscal Problems

Flat tariff structures in municipal services are responsible for under performance on revenue. There is no alteration in the general property tax systems and water supply prices. Low-income households are often at the regressive end of fiscal system. Improvements in tax revenues and consumer are willingness to pay if improvement in the institutions of service delivery.

Financial Problems

Urban bodies have limited financial resources systems and treasury management systems. As far as finance is concerned, the urban bodies depend on financial transfer from

Table 2.3: Gross Capital Formation in Infrastructure (at current prices)

(Rs Crore)

Infrastructure	*93-94*	*93-98*	*98-99*	*99-00*	*00-01*	*01-02*	*02-03*
Gross Capital Formation	45940	62903	70762	75120	94226	91050	90890
1. Electricity, Gas and water	23170	30756	36129	34255	36907	39180	42051
2. Transport, Storage, Comm. (2.1+2.2+2.3+2.4)	22770	32147	34633	40865	57319	51870	48839
2.1 Railways	5580	5069	5019	5307	5491	6981	9470
2.2 Transport by other	11304	16460	18153	21272	25802	21019	19028
2.3. Storage	136	456	599	353	1362	1525	1442
2.4. Communication	5750	10162	10862	13933	24664	22345	18859

Source: CSO

* includes roads, civil aviation, seaports and inland water.

other tier, of government or central government intermediation e.g. HUDC. Private financial systems are not ready to come forward to provide finance to social programs and safety nets for the poor. Urban local government across India is generally considered to be in extremely poor financial health (Kandwalla 1999).

Investment in Infrastructure

A huge amount of investments are required to bring India's infrastructure up to World standards. The details of the Gross Capital Formation in Infrastructure are explained in Table 2.3. Gross capital formation in infrastructure investment at current prices increased from Rs. 45940 crore in 1993-94 to 90890 crores in 2002-03. However it can be expressed that as a proportion of GDP to total investments in infrastructure declined from 5.4 percent in 1993-94 to 3.7 percent in 2002.03.

Urban Water Supply in India

In most developing countries, three most important environmental health problems that affect a large majority of population are contaminated water supply, inadequate sanitation and untreated solid wastes. The Global Water Supply and Sanitation Assessment Report states that at the beginning of 2000, one-sixth of the world's population was without access to improved water supply and two-fifth lacked access to improved sanitation (WHO/UNICEF 2000) In the absence of proper sanitation, people suffered from high levels of infectious diseases leading to high incidences of morbidity and mortality. This directly affected the ability of a country to maintain an efficient economy and implied great personal suffering among infected individuals and their families. More than one third of world population (about 2.4 billion people) lacked access to adequate sanitation facilities and four out of five of these unserved people lived in Asia (Cairncross 2003). Inadequate sanitation like unsafe disposal of human excreta, open defecation, lack of infrastructure (sewerage, drainage/sullage), and absence of hygiene management constitute a major threat to the health of the people. Thus, improving environmental health is the most cost-effective means of enhancing people's health and welfare. Nearly 1.7 million deaths every year is attributable to inadequate access to water, sanitation and hygiene (WHO 2002). Improving urban environmental health can prevent up to 44 per cent of the burden of the disease in the cities of the developing world. The problem of poor sanitation will become more serious in future as the city population continues to grow rapidly. It is projected that by the year 2030, the urban population will grow to 3.3 billion, over 90 per cent of which will accrue to the cities of developing countries. In the past ten years, diarrhoea has killed more children than all those lost to arm conflicts in almost sixty years since the Second World War. A child dies every 15 seconds from diarrhoea, caused largely by poor sanitation and unsafe water. Despite the efforts and investment many low-income countries continue to suffer from inadequate and unsafe sanitation.

The Millennium Development Goal advanced by the international community states that the aim is to halve the number of people without adequate sanitation facilities by 2015. To achieve this target, on an average every day, 350,000 people will have to gain access to improved sanitation between now and 2015 (IRC 2003). Despite the global commitments, the improvements made by many countries during the last one decade are very poor and the constraints identified are: financial difficulties, institutional problems, inadequate human resources, lack of political commitment, insufficient community involvement, inadequate operation and maintenance, lack of hygiene education, poor water quality, people's attitudes towards sanitation and insufficient information and communication (WHO/UNICEF 2000).

In India, utilizable water availability per capita per year is estimated at 1250 m3 (CWC, 1999)). Acute water shortages occur in many parts of the country due to unequal distribution of water resources over the year. The main reason is inadequate availability of water. Provision of safe drinking water to all is a dram of poor. According to National Water Policy, as 78 percent of rural and 85 percent of urban people had access to safe potable water. But in India 200 million people do not have access to safe drinking water at present and nearly 1.5 million children fewer than five die each year due to water borne diseases.

The renewable water resources amount to more than 200 cubic metres per capita in India, which is adequate to meet demand. However, water resources are not evenly distributed throughout the country, when North-East of India possesses abundance of water whereas the Western and Southern regions face deficiency of water. In India, some study estimates that with in a few decades availability of water will be about 170 to 200 cubic metres per person as against the world average of 5000 to 9000 cubic metres per person. The fresh water availability is explained Table 2.4. It is observed that out of 8649m3 of fresh per capita water availability in World, 1878m3 of fresh per capita water used in India.

Table 2.4: Fresh water availability

Fresh Water	*World*	*India*
Fresh water resources per capita (m3)	8,649	1,878
Fresh water withdrawal total percentage	71	26.2
Access to an improved water source (% of total population)	81	84
Rural (% of rural population)	71	79
Urban (% of Urban population)	94	95

Source: The World Bank The little green Data Book, from the world Development indicators – 2003.

WATER DEMAND FOR DIFFERENT USES

Domestic Water Supply

In India nearly 85 percent of the population depends on ground water as a source of water for domestic purposes. Some of the urban and rural population also uses surface water after conventional treatment for domestic purposes. According to a survey conducted by the Central Pollution Control Board nearly 28000 million litres of water per day is supplied to class II cities and I. The demands for different uses of water in India are given in Table 2.5. According to Ministry of Water Resources in India nearly 25 billion cubic metre (BCM) of water used for domestic purpose in India in 1990 and it is expected that the water demand for domestic purpose will increase 52 BCM in 2025 and 60 BCM in 2050.

Irrigation water supply

In India, irrigation is the most important source of water. Nearly 84 percent of water is used only for irrigation. The total water used for irrigation was estimated as 460 BCM in 2001 and it is expected that the irrigation demand for water will be increased to 770 BCM in 2025 and 800 BCM in 2050.

Industrial use and Power Generation

A large amount of water is used for industrial purposes as per the estimation of CPLB about 15 BCM of water was used in the industries as process water and cooling water in 2001. The future demand for industry will be 123 BCM in 2025 and 130 BCM in 2050. The water is also used for power generation; it estimated that the demand for water of power generation was

Table 2.5: Water Demand for Different Uses in India (BCM)

Different uses	*2001*	*2025*	*2050*
Irrigation	460	770	800
Drinking& Domestic	25	52	60
Industry	15	120	130
Power generation	19	71	120
Others	33	37	40

Source: - CPLB 2002.

19 BCM in 2001 and it will be increased to 71 BCM in 2025 and 120 BCM in 2050.

In most of the urban areas in India water supply is intermittent and consumers experience regular shortages with only a few hours supply of water on average per day acute shortages of water supply largely due to an unfair water distribution network. It is noted that higher proportion of water supply where the people live in adjacent of water distribution network and lower proportion in the areas where poor people live in far away from the distribution network. Due to lack of foresight and bad planning, 20 per cent of water has been wasted due to leaks. Inadequate financial resources are also responsible for the acute water supply problem in Indian cities. According to the World Bank estimate, there is an annual loss of Rs. 360 billion in India due to poor quality of water and sanitation. A staggering 90 million people in Indian villages do not access to safe drinking water and water on tap for 24 hours a day has been unheard of for decades in most Indian cities (ADB 1993).

Main Sources of Water in Urban India

Most of the households depend on their own sources such as tube well and boreholes for their day-to-day water requirements. The details of the main sources of water in urban India are furnished in Table 2.6. If over extracts water from the ground, the ground water tables are falling dramatic rates. A large number of the households in urban India depend on the piped water supply for their daily requirements. According to the India Infrastructure Report (2004) more than 70 per cent of the households depend on tap water and an insignificant number on tankers. Next to the tap water supply, access to underground water is the most important source of supply to the households such as wells, tube and hand pumps. Around the 22 percent of water is drawn from the tube where as 6.7 percent of water is extracted from the wells for the household purposes. Other sources that include tanks, ponds, springs rivers canals etc are insignificant in urban areas. Table 2.7 reveals the distance from principal source of drinking water. It is noted that 38.6 per cent of the urban households is collected water for drinking and other

Table 2.6: Main Sources of Water in Urban India

Sources	*No. of HH's (million)*	*Percent*
Tap	33.3	70.1
Tube wells	10.2	21.4
Wells	3.2	6.7
Tank/pond Reserved	0.1	0.2
Other Tanks/ponds for DW	0.0	0.1
River, canal, lake	0.1	0.2
Spring	0.0	0.1
Tanker	0.5	1.
Others	0.1	0.2
Not available	0.1	0.1
Total	47.6	100.0

Source:India Infrastructure Report 2004.

Table 2.7: Distance from Principal Source of Drinking Water: All Urban Households

Distance	*No. of HH's (million)*	*Percent*
Dwelling	18.4	38.6
Premises	12.9	27.1
< 0.2 km	15.00	31.5
0.2-05	0.80	1.7
0.5 – 1.0	0.4	0.7
1.0 – 1.6	0.1	0.3
> 1.6 km	0.0	0.1
Not available	0.0	0.1
Total	47.6	100.0

Source: India Infrastructure Report 2004.
Note: Te term 'Premises' means outside the dwelling but within the dwelling compound

hygienic purposes within their dwelling and around 21.4 per cent of the households within their premises. Nearly 31.5 per cent of the households in urban areas travel less than 0.2km distance to collect water for their day-to-day requirements whereas the 1.7 per cent of the people travels 0.2 to 0.5 km distance for water requirements. The remaining households are more than 0.5kms away from the source of water and that is the maximum distance travelled by the people to collect water.

WATER POLICY IN INDIA

The 74th constitutional Amendment in India provides locally elected Municipal Governments with the financial and administrative means to given the urban services. The 74th Constitutional Amendment has been ratified by the States Governments, which indicates that there is widespread understanding of the roots of current urban malaise in India.

There is also a growing consensus that political, administrative and financial reforms must be undertaken to enable progress to be made. The functional responsibilities of the municipalities, viz. water supply for domestic, industrial and commercial purposes and public health, sanitation and solid waste management are the prime responsibilities of the municipalities. Despite the constitutional empowerment there have been gross under performance, low quality of service delivery and lack of operating efficiency in many of the municipalities. The approach paper to the Ninth Plan estimates that 85 percent of India's Urban population has access to water supply and 48 percent to sanitation services. The accelerated urban water supply programme was launched in 1993-94. It aims at providing water supply to towns with a population less than 20,000 as per 1991 census. 2151 towns qualify for consideration under this scheme as per this criterion. As on December 31, 2002, 736 schemes (223 schemes sanctioned during ninth plan and 76 schemes sanctioned during 2002-03 have been approved at an estimated cost of Rs. 951.16 crore. As against this, Government of India has released Rs. 435.54 crore and the state governments have released Rs.311.41 crores during this period. The 10th plan envisages the provision of potable drinking water to every settlement in the country on a sustainable basis and pursuit of all possible measure for the rapid expansion and improvement of sanitation facilities in rural and urban areas (India Economic Survey 2002-2003).

The Central Public Health and Environmental Engineering Organization (CPHEEO), Government of India set the minimum water requirement as 125 to 200 lpcd (litres per capita per day) for cities with population of 50,000 and above. However, the Sakaria committee has suggested that per capital supply of 151.5 to 270.0 litres per day per head would be ideal goal for cities within a population of 1,00,000 and above. The National Master Plan of India has suggested that a water standard of 70 to 250 lpcd with an average supply of 140 1pcd irrespective of population size. According to (NIUA 2000) the estimate of the percentage of population covered under safe drinking water supply in India was 84 lpcd.

WATER ALLOCATION IN INDIA

At present, there are no legal or administrative frameworks, which allow for the reallocation of water from low value agricultural use to high value Urban and industrial use. India will face a severe water crisis if the available fresh water resources are not managed properly.

STATE WISE URBANIZATION AND WATER SUPPLY COVERAGE

Table 2.8 indicates the percentage of population covered with drinking water facilities in India. According to Economic Survey Report, in 1985, 72 per cent of the urban population covered with drinking water facilities in India but it was

Table 2.8: Percentage of Population Covered with Drinking water facilities in India

Water supply	*1985*	*1990*	*1995*	*1997*	*2001*
Rural	56.30	73.90	82.80	86.74	73.00
Urban	72.90	83.80	84.33	85.00	90.00

Source: Economic Survey, 2006-07 G01, New Delhi.

Table 2.9: State wise Urban Population and Percentage of Population covered by Water supply in India

States	*Percentage of Urban Population*				*% age of population covered by water supply*		
	1971	*1981*	*1991*	*2001*	*1992*	*1993*	*2001*
Andhra Pradesh	19.31	23.25	26.84	27.08	78.00	83.00	134
AP	3.70	6.32	12.21	20.41	100.00	100.00	100.00
Assam	8.82	9.88	11.09	12.72	40.00	11.00	30
Bihar	10.00	12.46	13.17	10.47	70.00	80.00	61
Chattisgarh	NA	Na	Na	20.08	NA	NA	NA
Delhi	89.70	92.84	89.93	93.01	99.00	90.00	NA
Goa	26.44	32.46	41.02	49.77	92.00	88.00	NA
Gujarat	28.08	31.08	34.40	37.35	98.00	98.00	133

(Contd.)

(Contd. Table 2.9)

States	Percentage of Urban Population				% age of population covered by water supply		
	1971	1981	1991	2001	1992	1993	2001
Hariyana	17.66	21.96	24.79	29.00	100.00	100.00	100.00
HP	6.99	7.72	8.70	9.79	100.00	100.00	100.00
J and K	18.59	21.05	22.76	24.88	98.00	100.00	100.00
Jharkhand	NA	NA	NA	22.25	NA	NA	NA
Karnataka	24.31	28.91	30.91	33.98	96.00	94.00	108
Kerala	16.24	18.78	26.44	25.97	75.00	6100	106
MP	16.29	20.31	23.21	26.67	88.00	90.00	185
Maharastra	31.17	35.03	38.73	42.40	98.00	98.00	175
Manipur	13.19	26.44	27.69	23.88	71.00	84.00	NA
Meghalaya	14.55	18.03	18.69	19.63	100.00	72.00	57
Mizoram	11.36	25.17	46.20	49.50	79.00	57.00	NA

(Contd.)

(Contd. Table 2.9)

States	Percentage of Urban Population				% age of population covered by water supply		
	1971	1981	1991	2001	1992	1993	2001
Nagaland	9.95	15.54	17.28	17.74	63.00	100.00	100.00
Orissa	8.41	11.82	13.43	14.97	50.00	53.00	239
Punjab	23.73	27.72	29.72	33.95	71.00	56.00	170
Rajasthan	17.63	20.93	22.88	23.38	100.00	100.00	100.00
Sikkim	9.37	16.23	9.12	11.10	74.00	62.00	NA
Tamil Nadu	30.26	32.98	34.20	43.86	49.00	49.00	94
Tripura	10.43	10.98	15.26	17.02	53.00	53.00	251
Uttar Pradesh	14.02	18.01	19.89	20.78	97.00	95.00	192

(Contd.)

(Contd. Table 2.9)

States	Percentage of Urban Population				% age of population covered by water supply		
	1971	1981	1991	2001	1992	1993	2001
Uttaranchal	NA	NA	NA	25.59	NA	NA	
West Bengal	24.75	26.49	27.39	28.03	80.00	86.00	106
A and N	22.77	26.36	26.80	32.67	86.00	86.00	
Chandigarh	90.55	93.60	89.69	89.78	100.00	100.00	100.00
D and Ni	0.00	6.67	8.47	22.89	100.00	93.00	
D and D	-	-	46.86	36.26	100.00	100.00	100.00
Lashadweep	0.00	46.86	56.29	44.47	-	100.00	100.00
Pondicherry	42.04	52.32	64.05	66.57	95.00	100.00	100.00
India	20.22	23.73	25.72	27.78	84.65	84.00	90 142

Source: Census of India – 1981, 1991 and 2001 and Annual Report: 1992-93, Ministry of Urban Development India.

increased to 90 per cent in 2001. It is noted that 90 per cent of the urban population covered with the drinking water facilities in India whereas the only the 85 per cent of the population covered in rural areas in India. State wise Urban Population and Percentage of Population covered by Water supply in India are given in Table 2.9.

The National Capital Territory of Delhi is the most urbanized that it has 93 percent population among all the States and Union Territories followed by Chandigarh (89.9 percent) Pondicherry (66.6 percent) as per census 2001. Tamil Nadu is the most urbanized state among the major states in India with 43.9 percent of the population living in urban areas followed by Maharastra (42.4 percent) and Gujarat (37.4 percent). The percent of urban population is the lowest in Bihar with 10.5 percent followed by Assam (12.7 percent) and Orissa (14.9 percent) meanwhile Himachal Pradesh is the least urbanised State that it has 9.8 percent of total population among all the states and UTs in India. According to Government of India estimates, by 2011 and 2021 India's urban population is estimated to be 366 million and 459 million respectively.

GROWTH OF METRO CITIES AND WATER SUPPLY ACCESS

As per census of India, the total number of cities in 1951 was only 5 and it increased to 12 in 1981. The number cities have been increased to 34 with population of 27.8 million in 2001 and it was a share of 37.8 per cent of total urban population of 107.9 million. At present, there are six mega cities in India, viz, Greater Mumbai, Kolkatta, Chennai, Delhi, Bangalore and Hydrabad. By the year 2021, it is expected that India will have the greatest Concentration of Mega cities in Asia.

INVESTMENT IN WATER SUPPLY

For water supply and sanitation, the investment need to clear the backlog is in the range of US$ 20-42.9 billion (India

Table 2.10: Percentage of Public Sector Investment in Water Supply and Sanitation during the plan period in India

Plan Periods	*Urban as $ of total Plan Expenditure*	*Rural as % of total Plan Expenditure*	*Percentage of Total Plan Expenditure*
First Five year – 1951-56	0.41	0.15	0.56
Second Five year – 1956-61	1.00	0.75	1.79
Third Five year – 1961-66	2.45	0.81	3.26
Fourth Five year – 1969-74	2.20	1.40	3.60
Fifth Five year – 1974-79	1.10	0.90	2.00
Sixth Five year – 1981-85	1.40	2.48	3.80
Seventh Five year – 1985-90	1.30	2.10	3.40
Annual Plan – 1990-92	1.35	2.00	3.35
Eight Five year – 1992-97	1.45	2.47	3.92
Ninth Five year – 1997-01	2.16	2.43	4.59
Tenth Five year -02-07	1.20	1.60	2.89

Source: Planning Commission of India.

infrastructure Report 1991). The details of the percentage of public sector investment in water supply and sanitation during the plan period in India are provided in Table 2.10. The additional investment to keep the services on track has been estimated by Indian planning commission to be in the range of US$ 2.5 to 3.7 billion for the period of 1996– 2001.

An estimate by the Society for Development Studies puts the investment requirements to clear the sanitation backlog up to 1995 at US$ 15.1 billion. Additional investment requirements for 1996-2002 for sewerage range between US$ 2.5 – 2.8 billion. The projections of investments are US$ 6 billion between 2001 and 2011 and US$ 6.5 billion for 2011-2025.

Urban water supply is state subject, maintaining capital assets and fixing and collecting water tariffs is the responsibility of the respective local bodies. But most of the urban local bodies are found to be weak, financially. The urban local bodies mobilize revenue from the internal and external sources. Internal sources include water tax, drainage tax etc., and non-tax sources include water charges, drainage charges etc. The external sources include loans and grants from the state government and international financial institutions. There has been a significant increase in percentage of total plan expenditure from 0.56 percent in the First Five Year Plan to 4.59 percent in the Ninth Five Year Plan but it has declined to 2.89 per cent in Tenth Five Year Plan. However the percentage of plan outlay for rural water supply has been more than the Fifth Five Year Plan.

MAJOR PROBLEMS IN THE WATER SUPPLY SECTOR

Water supply system in Indian cities has expanded in an unplanned manner. Most of the cities face acute water scarcity because, the situation of demand always exceed the supply. This leads to drastic depletion of ground water. Planners project future demand without considering whether the available supplies could be sustainable over a long period. Institutions, laws, government agencies, planning and engineering are concentrated only on the supply side management.

The conventional approach of supply side management may work only when the water is abundant and it is not suitable to a period of water scarcity and raising water demand. The following factors are responsible for weakening of the water supply sector.

HIGH SYSTEM LOSSES

Indian cities face a major water supply problem of leakage losses, which is termed as Unaccounted – for – water (UAFW). It affects, even the water quality. Many studies in India revealed that 50 to 60 percent of the water treated and pumped by the utility is lost or unaccounted for (Everything about water 2002). It has a direct impact on revenue generation and sustainability of the entire water supply system. The main causes for loss of water are old pipelines, poor quality of material and poor operation and maintenance. The classic problems of operation and maintenance are compounded by the problems of administrative management. According to un-habitat (2003), the percentage of unaccounted water is 50 percent in Kolkatta, 20 percent in Chennai, 26 percent in Delhi and 18 percent in Mumbai and 17 percent for Bangalore. Instead of going for 50 percent augmentation, water supply agencies should invest in leakage prevention, which could save additional investments.

TARIFF

The water tariff in highly subsidized in India. The willingness to pay concept is almost absent. The water production requires a huge amount of investments whereas water prices reflect neither the capital cost or production and maintenance costs. In most cases cross – subsidies from industrial consumers to households are prevalent. Further the subsidized rates encourage inefficiency and wastage.

ABSENCE OF SEASONAL TARIFFS

Every year during the summer, water utility faces serious water problem because water supply is limited and the demand for water is very high. However, the cities charge uniform or flat rate tariff system, throughout the year. Therefore, water conservation during the summer months is non-existent, leading to a high demand low return situation.

ABSENCE OF TELESCOPIC CHARGES

Most of the Indian cities charge water tariff rates on the basis of human consumption tool for fixing water tariff is on water tax as the percentage of the property. This tends the consumers to use any amount of water since the cost in any case is the same. Some institution such as Bombay Water Supply and Sewerage Board (BWSSB) has taken some initiative to review water rates so as to recover costs on a no gain and no loss basis. The national conference on water Management organized by Conference of Indian Industry in 1999 made a call for raising water charges to meet operation and maintenance expenditure. But many of the municipalities have not been able to revise water rates, as there was considerable opposition to do the same.

India's Infrastructure Report (2001) pointed out some of the major drawbacks in the existing tariff structure in water supply sector in India. There are:

1. There is only partial recovery of capital costs.

2. Tariff determined as per existing practices is largely aimed at recovering historical costs rather than meeting long – term incremental costs and increased investment, of an expanding system.

3. Tariff structure is not strictly followed due to political interference.

4. Attention is not focused on the affordability of and willingness to pay.

5. Tariff is not linked to the quality of service

6. Minimum lifetime rates are not defined.

This leads to a further depletion of municipal finances that depend strongly on loans and grains from the state. As it is municipalities have very restricted access to financial markets. Both at the city and state level, the lack of coordination between agencies responsible for infrastructure and urban development hinder progress.

A study conducted by the National institute of Urban Affairs (NIUA) on Urban water supply and sanitation showed that the costs of water provision were in excess of recoveries in nearly 76 percent of cities and, it means that expenditure exceeds the revenue of municipalities and the operation and maintenance costs of water supply systems were approximately 22 per cent higher than the receipts from water charges and water tax levied in lieu of water charges. The same study showed that the annual deficits on account of water (average annual per capital expenditure minus average annual per capital revenue) were Rs.20 per capital in metropolitan cities, Rs.40 per capital in cities in the population range of 100,000 and one million, and Rs.30 per capital in towns which have a population of over 50,000 but less than 100,000 persons. The survey has shown that the annual losses on operation and maintenance of the urban water supply systems would anywhere between Rs.9000 – 10,000 million. The major problems in the water supply sector are responsible for losses on revenue in the urban utilities.

THE MUNICIPAL WATER SUPPLY REFORM AGENDA

In India public services remain under state subject while the Central Government gives financial aids and directions to implement the public service scheme. Recently each state started water supply reforms agenda. The key elements of the reforms agenda comprise mobilization of financial resources, access to modern technology, Foreign Direct Investment and establishing a legal framework. In order to mobilize resources, authorities require regulator and facilitator to implement the reforms and to improve the information system by getting reliable data. According to the Ministry of Urban Affairs and Employment, the need for greater involvement of the private sector and local community participations has been encouraged in the water supply sector also. According to India water Resource management (1994), the key elements of the reforms of the institutional framework are:

1. Democratic decentralization of responsibility for urban water supply services to municipalities.

2. Business like provision of urban water supply services through commercialization of existing providers. Private sector participation and contracting out of operations and

3. Introduction of mechanism to ensure customer orientation of urban water supply agenda, including in providing service to disadvantaged groups.

The 74th Constitutional Amendment provides locally elected municipal government with the financial and administrative means to govern effectively. The Constitutional Amendment laid the foundation for democratic decentralization by giving municipalities the power of self-rule. The importance of local democracy ensures the provisions of transparency, responsiveness and efficiency. The functional responsibilities of the municipalities are left for the states, but states are strongly encouraged to reinforce the power of elected municipal members. The participation of elected municipal members is encouraged in development, financial and investment projects in water supply. The main objective of reforms in municipal water supply is to promote bottom up approach to development.

The State Finance commission helps every municipality in India in order to increase their financial conditions. The internal resources could be transferred into well-established municipalities from weak municipalities. There is a co-ordination between the states and municipalities in the implementation of programmes. This leads to raise their resources and help to increasing financial resource. This means that greater freedom to municipalities in fixing taxes and tariffs. However, state governments are not being able to implement the local urban policies.

In many of the developing countries drinking water is provided by the public sector, since it is a typical case public good. Generally the public utility takes charge of the entire cycle of water production, distribution and management through a vertically integrated monopoly. Recently the changes of reform in India where these functions are partially or fully assigned to private sector through a franchise bidding process (coing et al, 1989, Silva et al. 1998 Coing H. de Lara P and Montano 1989). The municipal Act does not prohibit the use of any person for delivering specific services. The Hyderabad Municipal Corporation Act 1955 (No.11 of 1956) for example, clearly states that the commissioner when authorized by the corporation can enter into an arrangement with any person for the supply of water. So, the private sector involvement is a part of the reforms in water supply services.

Some estimates reveals that the development of financial markets could further develop water supply projects, and improving operational efficiencies and reducing losses, are important causes for the involvement of private sector. The private sector has already become involved through the development of water tankers, and bottle water corporations.

The Finance Commission of States (Kerala, Mathaya Pradesh, Punjab and Tamil Nadu) has recommended contracting out of certain municipal services to private agencies. Experiences in India and internationally have shown that involving the private sector in Urban water supply provision can further boost efficiency and service standard (Ministry of Urban Affairs and Employment, India). The role of public private participation (PPP) in water supply to widely studied and debated (NIUA 1997). The PPP study constructed by India infrastructure Report (2004) that widely discussed the emerging private sector involvement in water supply sector.

WATER IN INDIA

The water supply and sanitation sector will face enormous challenges over the coming decades. In India, the groundwater is consumed directly without any sort of treatment and disinfection. Its quality is therefore a cause of concern. The national objectives of reducing morbidity and mortality largely depend on the reduction of diarrhoea, jaundice, etc. In fact, no water supply and sanitation programme can be successful if water-related illnesses are not reduced. It is a matter of concern that despite the progress made with water supply, the level of water-related illnesses continues to be high. Approximately 10 million cases of diarrhoea, more than 7.2 lakh typhoid cases, and 1.5 lakh viral hepatitis cases occur every year. A majority of them are contributed by unclean water supply and poor sanitation. Micro-level studies revealed that availability of clean water; sanitation, and hygiene interventions reduce diarrhoeal diseases on average by between one-quarter and one-third. The details of cases and deaths due to water borne diseases in various states are furnished in Table 2.11.

Causes of contamination of water are indiscriminate use of chemical fertilizers and chemicals, poor hygienic environment of the water sources, improper disposal of sewage and solid waste, pollution from untreated industrial effluents, and over-

Table 2.11: Cases and Deaths due to Water-borne Diseases in Various States, 2006

States	*Diarrhoeal Disease*		*Viral Hepatitis*		*Typhoid*	
	Cases	*Deaths*	*Cases*	*Deaths*	*Cases*	*Deaths*
AP	1215659	124	17846	28	135550	12
Arunachal P	32032	30	553	6	9098	23
Assam	..	..	..	..	..	..
Bihar	..	..	..	..	..	..
Chhattisgarh	95202	13	1491	2	21474	6
Goa	7631	0	15	0	68	0
Gujarat	382056	4	9396	16	7290	0
Haryana	285342	42	3983	11	5688	4
HP	347055	28	835	11	26327	5
J&K	519317	32	5882	0	42369	0
Jharkhand	14752	1	51	0	4707	284
Karnataka	939221	1279	14980	24	96147	5
Kerala	475510	4	7018	6	6219	2
MP	318935	88	2499	9	28654	29
Maharashtra	695723	93	43215	131	39663	8
Manipur	13614	17	346	0	2421	2

(Contd.)

(Contd. Table 2.11)

States	Diarrhoeal Disease		Viral Hepatitis		Typhoid	
	Cases	Deaths	Cases	Deaths	Cases	Deaths
Meghalaya	178260	33	294	2	6709	1
Mizoram	18063	20	546	11	1392	2
Nagaland	9176	0	112	0	2328	0
Orissa	373748	40	2687	38	15387	9
Punjab	182451	64	3829	17	17008	3
Rajasthan	318169	21	3869	78	14084	131
Sikkim	51433	8	290	2	428	2
Tamil Nadu	116062	12	4523	0	36973	0
Tripura	150750	47	2520	14	18547	19
Uttarakhand	94746	6	3381	0	15020	2
UP	284709	55	3716	6	42648	13
WB	2622968	964	7433	205	110835	70
A&N Islands	22752	2	213	4	3055	0
Chandigarh	..	..	..	..	..	..
D&N Haveli	74661	4	126	3	646	0
Daman & Diu	109	0	3	0	33	0
Delhi	94398	85	4080	42	13774	18
Lakshadweep	7316	0	86	0	6	0
Pondicherry	137443	8	615	7	1936	1
Total	10079263	3124	146433	673	726484	651

Source: National Health Profile 2006, CBHI, DGHS—MoHFW.
Note: '..' means not reported.

Table 2.12: Burden of Major Communicable Diseases in Various States, 2006

States	*Malaria*		*TB*	
	Cases	*Deaths*	*Cases*	*Deaths*
Andhra Pradesh	39099	0	142057	1184
Arunachal Pradesh	31215	0	1593	38
Assam	67885	113	..	..
Bihar	2733	1	..	..
Chhattisgarh	187950	3	8689	12
Goa	3747	1	2228	0
Gujarat	179023	54	41730	238
Haryana	33262	0	29900	227
Himachal Pradesh	129	0	14705	140
J&K	268	0	2346	28

(Contd.)

(Contd. Table 2.12)

States	*Malaria*		*TB*	
	Cases	*Deaths*	*Cases*	*Deaths*
Jharkhand	193144	21	15516	5
Karnataka	83181	26	76687	745
Kerala	2554	6	13840	181
MP	104317	44	18866	132
Maharashtra	47608	104	52998	795
Manipur	1844	3	482	20
Meghalaya	16816	41	1900	29
Mizoram	10741	74	936	28
Nagaland	2987	0	838	0
Orissa	396573	255	11443	178
Punjab	1883	0	20612	106
Rajasthan	52286	22	71180	695
Sikkim	69	0	2155	36

(Contd.)

(Contd. Table 2.12)

States	*Malaria*		*TB*	
	Cases	*Deaths*	*Cases*	*Deaths*
Tamil Nadu	39678	0	28979	73
Tripura	18008	20	971	27
Uttarakhand	1242	0	3522	8
UP	105303	0	109898	161
WB	185964	175	89276	820
A&N Islands	3954	0	2898	10
Chandigarh	432	0	..	..
D&N Haveli	1166	0	1544	36
Daman and Diu	104	0	25	3
Delhi	11330	135	44	993
Lakshadweep	0	0	39	0
Puducherry	44	0	7738	125
Total	1816342	963	789135	7073

Source: National Health Profile 2006, CBHI, DGHS—MoHFW.

Note: '..' means not reported.

exploitation leading to quality degradation. Thus, the supply of additional quantity of water by itself does not ensure good health, proper handling of water and prevention of contamination are also equally important. The details of burden of major communicable diseases in various states are provided in Table 2.12. It is found that the total number of cases have been registered (1816342 cases) during year of 2006 due to water related malaria diseases. Out of this, 963 cases have been lost their lives in overall India. It is noted that the death cases have been registered in Assam (113), Maharastra (103),Orissa (225), West Bengal (175) and Delhi (135). The study reveals that 789135 cases have registered due to important diseases like TB and out of these 7073 persons lost their life.

POSITION OF TAMIL NADU

Tamil Nadu is dependant on rainfall for its water supply. It has no perennial rivers and geologically too, the State is mostly covered by hard rock formations where the availability of ground water is uncertain. The population of the State as per 2001 census is 62.11 million out of which 27.24 million is urban and 34.87 million is rural population. The decadal growth stands at 11.19 per cent. With fast urbanisation and increasing expectancy of the people, the demand for water is on the increase but the available resource is scarce. This calls for more efficient use of resources and more reliable delivery system.

DRINKING WATER SUPPLY PROBLEM CHENNAI CITY

The drinking water requirements of Chennai city are met through a network of four surface water reservoirs namely Poondi, Cholavaram, Redhills and Chembarambakkam. But these sources are entirely dependant on rain and the city experiences acute water scarcity very often due to failure of monsoon. In order to augment the City water supply, ground water sources from the three well fields of A.K. Basin viz. Minjur, Panjetty and Tamaraipakkam which were developed by the

Public Works Department in 1968 were taken over by CMWSSB based on the recommendations of the UNDP study. However, all these sources became inadequate during the drought year 1982. Based on the recommendations of further study made by CMWSSB in collaboration with the UNDP, three more well fields were commissioned in the A.K. Basin, viz., Poondi, Kannigaipair, and Flood Plains in the year 1987. The originally designed capacity of the six well fields was intended to draw 180 mld. However, pumping by Metro water never reached such a quantity. A maximum of 120 mld was abstracted during 1987. Due to drought conditions and due to continuous drawal of ground water, the water table of the aquifer went down, year after year. During 1993-94 the total abstraction was just 45 to 55 mld from the Board wells in all the six well fields.

URBAN WATER SUPPLY PROGRAMME

An increasing urban population has been creating a huge gap between demand and supply of water every year. The last Census estimated that approximately only 70 per cent of urban towns have access to safe drinking water. The minimum per capita supply of water required in urban areas varies from 70 lit/day to 130 lit/day, and this requirement of water supply varies according to the land use classification of the towns. The sources of water supply, purification pumping of water, storage and distribution varies from place to place. In Tamil Nadu, out of 744 towns (including corporations and municipalities), 145 towns are not fully provided with water supply. The problem of drinking water is more acute in rural and urban town panchayats. 45 out of 370 urban town panchayats and 97 of the 241 rural town panchayats are not fully provided with water supply. About 15 per cent of the urban population is yet to be provided access to drinking water, highlighting the fact that the urban water supply in Tamil Nadu is far below the national average.

For instance, 86 out of 104 municipalities, all corporations, and 412 out of 611 town panchayats have water supply below the national average of 90 lpcd. With the provision of water supply augmentation schemes to 249 towns in India the total investment was Rs. 559.96 crores, during the Ninth Five Year Plan. During the year 2002-2003 water supply schemes to 72 towns were completed with an investment of Rs. 170.78 crores , and in the year 2003-2004, 60 towns were covered with an investment of Rs 134.06 crores. In respect of urban sector, at the beginning of the 10th Plan, 419 towns comprising 5 Corporations, 64 Municipalities and 350 Town Panchayats had inadequate water supply. During the year 2002-03 water supply augmentation schemes were completed for 72 towns. During 2003-04, 60 towns were provided with water supply augmentation scheme. For the current year it has been programmed to complete water supply improvements schemes for 50 towns. The remaining towns with inadequate water supply will be covered in the remaining period of the 10th Plan.

SCHEMES FOR SUSTAINABILITY OF SOURCES

Even though schemes are designed and implemented with per capita water supply as per the norms, over a period of time the quantum of supply gets diminished mainly due to depletion in the water potential. Hence to ensure the sustainability of the source of water supply schemes, recharge structures like check dams, percolation ponds, subsurface dykes, rain water harvesting structures are to be established. TWAD Board in association with Institute of Remote Sensing, Anna University carried out a recharge project and identified 13,357 suitable areas for construction of recharge structures. Out of 13,357 locations, recharge structures have been provided with 5,444 locations leaving a balance of 7,913 locations where recharge structures have to be provided. It is indicated that an amount of about Rs.180 crores would be required to create the structures in the balance locations. It is important to see that construction of check dams and percolation ponds do not conflict with the already existing tank systems. Since the tanks are the traditional water harvesting structure, setting up of recharge structures should complement water conservation aspect and not be based on water acquisition in the catchment areas.

Rain Water Harvesting has to be prioritized based on the hydrogeology and socioeconomic conditions. It has to be employed in a river basin or sub-basin where the water outflows unutilized either through the surface or through the ground water aquifer. Availability of excess surface flows (over and above the present use) must be a pre-condition for putting up any artificial recharge structure. In the case of locations where tanks are not significant, the water harvesting structures such as major check dams, minor check dams and percolation ponds should be given priority. Further the intensity of wells in the given location should also be given weightage in deciding the choice of water harvesting structures. Any structure that helps to recharge the wells in the zone of influence is more important. In some locations the soil types also decide whether the water harvesting structures should be for supplemental irrigation or for recharge only. Therefore, a master plan is needed for rainwater harvesting, basin wise. The techniques to be adopted for urban and rural areas are quite different.

RECYCLING OF WASTE WATER

Wastewater constitutes a major quantum of water unused. Water needs to be reused as many times as possible before it is let off. Recycling of wastewater would be a possible option in future of reusing the wastewater. This will also serve as a conservation mode of available water. Recycling is widely practiced in Israel, the Middle East, and Singapore and Oxford in UK. Producing potable water from sewage is cheaper than desalinating water. Besides sludge generated in the process can be sold as bio-fertiliser after due treatment. It is concluded that the provision of urban water supply in India are challenging task to every urban local governments. When there increase in the rate of growth of urban population, urban problems are also increasing at the rapidly. This chapter highlights the gap between urban service providers and the consumers. It is inferred that the role of government must be increase to provide the urban services to the people.

3

Urbanisation, Growth and Evolution of Water Supply System in Chennai City

This chapter focuses on the historical background of the development of drinking water sources in Chennai City. It sketches the origin, historical development, growth of the city, its population and urbanisation, along with the growth of its commercial and business activities, art and culture. Urban growth and development often contributes to water resource problem. This study also evaluates the performance of the Chennai Metro Water Supply and Sewerage Board (CMWSB).

Evolution and Development of Chennai City

Chennai City[1] was a small fishing village at the time of the arrival of East India Company. This village later grew into the present metropolitan city. The rapid expansion of urban settlements around Fort.St.George was due to trade between the East India Company and Indian merchants and craftsmen (CMWSSB, 2001).

Chennai covered an area of 27sq.miles (68sq.km) in 1763 with a population of nearly half a million. The Census of 1871 and 1881 gave the population of the then Chennai as 3, 97,552 and 4, 05,848 respectively. It was also a commercial and trading centre because of the Madras Port. The port traded with every part of the world, exporting coffee, cotton, etc and importing piece goods, iron ore and other metals. Chennai City was the centre on which all the military roads converged viz, Bangalore Road, the Calcutta Trunk Road and the Grand Southern Trunk Road. There were two-railway lines, one connecting Bangalore and Bombay and other connecting Tuticorin. The third railway line was the Northern line to Kolkatta. A system of canal navigation was developed along the eastern coast called the Buckingham Canal.

By the end of the nineteenth century, Madras City had acquired the 13th place in population among the cities of the world, third place in the sub-continent, after Calcutta and Bombay. The city is adorned with many public buildings and other important monuments of historical significance.

The style of architecture of many of the building is the adaptation of the Indo-Sarcenic. The suburbs of Chennai City were then dotted with a number of elegant houses, occupied by the Europeans and the rich Indian community.

Lanchester worked on a plan for Chennai City in 1916. Subsequently, Town Planning Schemes covered the Town Planning Act came into being in 1920. Thiyagarayanagar was developed as a residential suburb in 1930 on the beds of a big tank. At that period of time urban problems began to intensify. An enquiry made in 1933 reported that there were over 2 lakh persons, including 30 per cent slum population. The limits of Chennai City were extended in 1941 by the addition of Mambalam. The extent of the area of the city became 77.21sq.km in 1941. All the neighbouring areas of Chennai City were gradually merged into the city. The limits were further extended in 1946 adding surrounding villages from the Chengalpattu District to bring the extent of the city to 128.83 Sq.km (CMDA and TRF, 1991).

Chennai City after Independence

The period between 1947 and 2001 witnessed unprecedented growth in both area and population of Chennai City. The urbanised area increased from 130 Sq.km in 1947 to 172 Sq.km in 2001. This growth has been triggered by progressive industrialisation and exodus after Independence. Apart from several large industries in the central public sector, many private automobile and ancillary industries were also established. Two Industrial Estates were also developed under State Government's initiative. The Chennai harbour was widened and deepened to cater to the needs of increasing exports and imports.

Geography of Chennai City

Chennai City, the capital of Tamil Nadu, is a city of contrasts and diversities. It is located near the 13° North Parallel and 80° longitude; stretching 19km along the Coromandal Coast and extends about 9 km inland. The Kanchipuram and Tiruvallur Districts surround the city. The topography of Chennai City is very flat, the contours ranging from 2.0 to 10.0 meters above the sea level. Cooum and Adyar are the two rivers flowing through Chennai City, which are non-perennial, but have a substantial flow during the monsoon. Buckingham Canal and Ottery Nullah.are the two watercourses where Buckingham canal runs along the coast adjoining the Bay of Bengal and Ottery Nullah starts from Villivakkam tank and joins the Buckingham Canal at Basin Bridge (TNPCB, 2001).

There is uneven distribution of land utilisation pattern in Chennai City. The North and West are the most congested areas, while density of population is comparatively much less on the semi circular belt around these crowded areas. During 1981 and 2001, the settlement patterns at the outskirts of Chennai City limits are on the increase, and this is due to the fillip given by rapid growth of transport, and increasing suburban conglomeration.

The soil of Chennai City falls broadly into three types, viz., alluvial, laterite and red soil. Sandy coastal alluvium is found along the coastal areas; red soil is found in the North West and clay in the West and North of Chennai City. The aquifer of Chennai City can be broadly classified into hard crystallions and prong formations. In the hard crystallioness, the degree of weathering and the presence of rocks vary gradually. However, the general zones of weathering do not extend beyond 12.0 meters below ground level. The data on existing bore wells indicate that wells show a moderate yield of 40 to 150 litres per minutes from the fractures encountered at a depth between 25 meters and 50 meters [Kapil, Gupta, 2002].

Climate and Rainfall

Chennai City generally has hot climate, which can be termed as tropical maritime monsoon type. The temperature in Chennai City is always on the higher side. The minimum temperature ranges from 21°C to 24°C between December and February and maximum average temperature records 37°C in May. Mean relative humidity is high throughout the year, and it varies from 65 per cent in May-July to 80 per cent in October-December.

Chennai City being a coastal city gets a fairly high intensity of rainfall. The quantum of rainfall varies significantly in different years. With a mean annual temperature of 30°C, Chennai City belongs to tropical climate. The city gets rain from two monsoons. One is North-East Monsoon and the other is South-West monsoon. The South-West Monsoon brings rain between June and August; Chennai City's weather pattern is dominated by North-East monsoon, which normally occurs between October and December. Generally, during the North-East monsoon, rainfall is intense when low depression occurs in the Bay of Bengal and moves towards the East Coast. Storms and cyclones give considerable amount of rainfall during the North-East Monsoon.

The supply of drinking water in Chennai City depends more on surface water. The quantum of surface water depends on rainfall in the catchment areas of its reservoirs. Even a single failure of monsoon makes a big difference in the storage level of these reservoirs. History reveals that water supplied to the households in Chennai City becomes scanty in the year after monsoon failure. The average rainfall received during the South West and North East monsoon periods in Chennai City between 1978 and 2008 is shown in Table 3.1.

When compared to the average rainfall of 1440 mm in Chennai City, almost all the years since 1978 received rainfall below the average level. The lowest rainfall is recorded in 1999, whereas moderate rainfall were in the years 1983, 1990, 1991, 1995, 1996, 1997 and 2001. The inference is that the rainfall has not been sufficient to meet the growing water demand of Chennai City.

The drinking water for Chennai City depends to a larger extent on the surface water sources. The surface water in turn depends on rainfall levels every year. Normally rainwater reaches the banks and drains into the river systems. Thus, a very large quantity of rainfall is wasted into the sea. The remaining is lost as evaporation, seepage into the ground enriching the ground water, and so on. It is very difficult to estimate the surface run-off from the rainfall. It depends on several factors such as rainfall intensity, rainfall duration, and pre-rainfall condition of the catchments areas, humidity, soil nature and the slope of the rain fed land. The difference between

Table 3.1. Average Rainfall Received in Catchments Areas of Chennai City, 1978-2008

Year	*Rainfall in city Reservoirs (in mm)*
1978	1389
1979	1139
1980	1121
1981	1006
1982	859
1983	1404
1984	1417
1985	1109
1986	904
1987	886
1988	1063
1989	891
1990	1434
1991	1426
1992	848
1993	1345
1994	1282
1995	1566
1996	2123
1997	2169
1998	1334
1999	624
2000	1053
2001	1400
2002	1129
2003	810
2004	1171
2005	2208
2006	1240
2007	1449
2008	1655

Source: Annual Reports 2002 and 2008, CMWSSB, p.60.

the available supply of surface sources and the total water demand is met by exploiting ground water.

Table 3. 2. Area and Population for Chennai Metropolitan Area, (CMA) Chennai Urban Agglomeration (CUA) and Chennai City (CC)

Year	*CMA*			*CUA*			*CC*		
	Area (Sq. Kms)	*Popula tion (Lakhs)*	*Decadal Varia-tion (%)*	*Area (Sq.Kms)*	*Popula-tion (Lakhs)*	*Decadal Varia-tion (%)*	*Area (Sq. Kms)*	*Popula-tion (Lakhs)*	*Decadal Varia-tion (%)*
1961	1177.00	21.34	-	226.07	19.44	-	128.83	17.49	-
1971	1177.00	34.19	60.21	530.77	31.69	63.02	128.83	25.72	47.06
1981	1177.00	46.28	35.37	571.93	42.89	35.31	170.00	32.76	27.35
1991	1177.00	59.19	27.89	571.93	53.61	24.99	170.00	37.95	15.82
2001	1177.00	75.22	29.00	571.93	64.24	19.82	172.00	42.16	11.09

Source: Census of India, 1981, General Population Table 1991: Census (Provisional) Tamil Nadu, and Census of India 2001.

Urban Growth and Water Crisis

Increasing concentration of population in a specific land area is a concomitant process of urbanisation. Rapid growth of population and, therefore, high density of population in Chennai City have led to crisis in water supply. The area and population for Chennai Metropolitan Area (CMA), Chennai Urban Agglomeration (CUA) and Chennai City are presented in Table.3.2.

The population of Chennai Metropolitan Area (CMA), Chennai Urban Agglomeration (CUA) and Chennai City have been increasing over the years; however, the percentage decadal variation of population has been declining drastically after the year 1971.

Table 3.3 shows that Chennai City had a population of 3.97 lakhs in 1871, which increased to 42.16 lakh in 2001. The population of Chennai City has been on the increase since 1921. It remained almost unchanged up to the year 1921. In 1931, the area of the city increased to 75 Sq.km and further extended to 129 Sq.km in 1951. In 1978, adding 12 Panchayats in the outskirts of Chennai City further enlarged the city area. During 1871-2001 there was more than a nine fold increase in population. The pace of growth of population in Chennai City during 1981-1991 was not as rapid as during the three decades of 1951-1981. Nevertheless, the density of population has shown a steady and stable increase over the years, despite the enlargement of the City three fold. In the year 2002, the Chennai City had a population of about 45 lakhs.

The growth of population was more due to the increase in natural growth rate than a rise in the flow of migrants. In 1971-81 and in 1981-91, the decadal rate of growth of population was 24.42 and 17.02 per cent respectively. In 1991-2001, the rate of growth of population was 10.32, which is 6.7 percentages less than the rate of growth of population in the previous decade.

Table 3. 3. Area, Density, Size and Growth of Population in Chennai City, 1871-2001

Census year	Area in square km	Density of population (per Sq.km)	Index of growth of density	Total population (Nos)	Increase population		Index of growth
					Absolute	Percentage	
1871	71	5600	100	397552			100
1881	71	5700	102	405848	8296	2.00	102
1891	71	6400	114	452518	46670	10.30	114
1901	71	7800	139	509346	56828	11.20	128
1911	71	8100	145	518660	9314	1.80	130
1921	71	8300	148	526911	8251	1.60	132
1931	75	9800	175	647230	120319	18.60	162

(Contd.)

(Contd. Table 3.3)

Census year	*Area in square km*	*Density of population (per Sq.km)*	*Index of growth of density*	*Total population (Nos)*	*Increase population*		
					Absolute	*Percentage*	*Index of growth*
1941	75	11800	211	777481	130251	16.80	195
1951	129	11100	198	1416056	638575	45.10	356
1961	129	13600	243	1729213	313085	18.10	436
1971	129	19100	341	2469213	740072	30.00	622
1981	170	19100	339	3266034	797621	24.42	823
1991	170	22077	394	3822011	555977	17.02	961
2001	172	24231	433	4216268	394257	10.32	1061

Source: Complied from Annual Report, 1951, Health Department of the Corporation of Chennai City and Census of India 1968, 1971 and 2001.

Evolution and Development of Water Supply in Chennai City

The following section of the present chapter explains the evolution and development of water supply in Chennai City over a period of time. The rapid growth of population and the increased economic activities warrant a water supply much more than the present volume of supply. Since the water sources are limited, huge public investment has been made in different alternative sources at a higher marginal cost. Thus the development of water supply system in Chennai City has progressed in stages even prior to Independence. Historically, the native people of Chennai depended on a network of tanks, ponds, lakes, temple tanks and wells for their water needs. Well was attached to a house and the other sources of water were shared by households.

The first institutionalised public water work was executed in 1772 when Chennai was under the control of the English East India Company. The water supply system was designed to supply 0.635 million litres per day (mld), equivalent to 140000 gallons per day from a cluster of 10 wells to Fort.St.George (A. Vaidyanathan, 1997). The sources of water supply to Chennai City had been from shallow wells for many years, and it was first in the year 1866 that it was decided to have a public water supply scheme. In 1872, the scheme was executed by combining the water supply needs for Chennai City and to irrigate 3500 hectares of wasteland surrounding the city.

Development of Cholavaram and Redhills Lake

The Cholavaram and Red Hills tanks are centuries old, perhaps formed by Cholas and Pallavas for irrigation purposes. The bunds of these reservoirs were raised and strengthened at a cost of Rs.18.50 lakhs in 1872 to meet the water demands of a population of 0.47 million (Manual, Chennai Corporation, 1957, p.1). The reservoirs received water from the Kortalaiyar River for few days between the months of July and November usually from floodwater. This river water was diverted into the Red Hills lake by means of a masonry anicut, built across the river of Tamaraipakkam, about 17 miles North-West of Chennai City. The water diverted was first led into the Cholavaram Lake and then to the Red Hills Lake through the upper and lower supply channels. These two tanks met the water demands of Chennai City till 1900. Further developments took place only after 1907, which were the construction of an outlet tower and roughing filters at Red Hills, underground conduit canal to take water to the city and the making of slow sand filters at Kilpauk (Manual, Chennai Corporation, 1957. p.2).

Progress of Kilpauk Water Works

In 1881, Jones Tower constructed the intake tower at Red Hills, through which water was drawn and conveyed to a masonry shaft at Kilpuak through an open channel. Water was supplied to areas such as George Town, Walltax Road and the surrounding areas through a small network of distribution pipelines. But it failed. The Kilpauk Water Treatment Plant was constructed in 1914 during the British regime. Several major works were undertaken in 1914 to improve the existing mode of water supply, designed and executed by special engineer Madely, Corporation of Chennai. They include:

1. A covered underground masonry conduit 7 miles long, starting from Red Hills Lake and leading to Kilpauk water works, with a carrying capacity of about 23 million gallons of water per day.

2. Further 14 slow sand filters 200′′ x 100′′ to filter and purify the water at the Kilpauk water works.

3. Pure water tank filters the size of 150′ x 150′ each capable of storing about ½ a million gallons of water, filtered before being pumped into the overhead tank and the distribution system thereafter.

4. Three heavy duty steam pumping units.

5. A 48" size pumping main of about 3000 feet long from Kilpauk pumping station to the shaft, where the tanks main starts.

6. The remodelling, alteration and extension of the distribution system so as to ensure an adequate supply, under sufficient pressures at all points of the distribution system (Manual, Chennai Corporation, 1957, pp 2-3).

According to the estimates of Madeley, these improvements were considered adequate for meeting the water needs of the City till 1961. But due to the rapid growth of city and more than the expected levels of population growth it became apparent by 1930 that further improvements were necessary to provide adequate water supply to Chennai City.

Masonry Raw Water Conduit (I Conduit) was constructed of size 5′′ x 4′′ with a capacity to convey 104 mld in 1914. One Steel Overhead Tank with a capacity of 6.8 ml was constructed to act as balancing storage. One additional underground tank of 12.56 ml was constructed in the year 1923. Conduit II with a capacity of 146 ml and rapid sand filters of 45 ml for conveyance of raw water capacity with 6 numbers of filter beds, were constructed in 1955. The two underground tanks of 11.35 ml capacity and one overhead tank of capacity of 6.8 ml were constructed in 1957 and 1965 respectively. In 1977 one underground tank was constructed. In 1986 a new intake tower was constructed at Red Hills for drawing additional quantity of water. Conduit III of 190 mld capacity was constructed and commissioned in the year 1989. One underground tank of 11.68 mld capacity was constructed in the year 1990. In 1991 a new water treatment plant of capacity of 90 mld was constructed and commissioned with 12 numbers of filter beds. Apart from water treatment, this plant is also used for water distribution to the adjoining areas. The total capacity of the plant is 327 million litres per day. The raw water for treatment is drawn from the Red Hills Lake through 12-km long brick masonry conduits. The method of treatment is pre-chlorination and addition of chemicals such as alum, and lime followed by flocculation, chlorination and filtration through Rapid Gravity Filters. This treatment was slowly upgraded to meet the increased demand for 327 mld of water. In order to convey treated water to the north and south of Chennai City, High Tension and Low Tension Pump Houses have been provided.

Raw water from Red Hills Lake is pumped through a Raw Water Pump House. The capacity of the treatment plant is 300 million litres per day. A residual laboratory has been provided for assessing and confirming the quality of treated water; water samples are collected in each and every stage of treatment process and tested. In order to meet the increased water demand, a 530 mld water treatment plant is being constructed on the bank of the Chembarambakkam Lake to enable treatment and supply of water to Chennai City. The proposed technology in treatment is pulsator technology (Annual Report, 2001-02, CMWSSB.p.43). To meet the growing water demand for more than one million population, a reservoir was constructed at Poondi across the Koratallaiyar River, known as Poondi reservoir.

Establishment of Poondi Reservoir

The Poondi reservoir has acquired a special status when the city prepared to receive Krishna water. The reservoir served the much-felt need of storing the floodwater across Koratalliyar upstream of Tamaraipakkam Anicut. The reservoir built at a cost of Rs.6.5 lakhs was declared open in 1944. It raised the total surface storage capacity from 100 mcm to 180 mcm. The water impounded in the reservoir is drawn into Red Hills Lake, through the upper and lower supply channels. The capacity of this reservoir is 2573 million cubic feet equal to the capacity of

Table 3.4. Characteristics of Water Reservoirs of Chennai City

Reservoirs	*Maximum water elevation (metre)*	*Draw of Elevation (metre)*	*Maximum usable volume (million cu.mt)*	*Full area surface (million sq.mt)*	*Mean depth (metre)*	*Catchments area (sq.km)*
Poondi	42.07	33.56	76.7	34.3	2.2	2950
Cholavaram	19.66	14.24	22.5	7.0	3.4	487*
Red Hills	14.69	8.89	71.0	20.6	3.8	76
Total			170.2			3513

Source: Madras – 2011, policy imperatives and agenda for action CMDA, 1991.

* Includes Tamaraipakkam drainage area.

both Red Hills and Cholavaram Lakes (Manual, Chennai Corporation, 1957. p 3-4).

Supply of water for irrigation from the Red Hills Lake ceased during the 1940s and 1950s. Subsequently, the treatment capacity of water was increased at Kilpauk by the provision of rapid-gravity filters, with a capacity of 45 mld, which was later increased to 135 mld and then to 190 mld. In 1957-58, the UNDP had suggested that the ground water in the Araniyar and Kortalaiyar basins be used for the drinking water requirements instead of costly desalination plants (Down to Earth, 2004). When UNDP started drilling bore wells, the farmers realised the importance of commercialisation of groundwater [R. Sakthivadivel, 2003, p.10]. Thereafter many efforts were made to tap the ground water instead of increasing the storage in reservoirs.

The features of Poondi, Cholavaram and Red hills reservoirs supplying water to Chennai City are given Table 3.4.

The total surface storage capacity of water available in Poondi, Cholavaram and Red Hills reservoirs is 170.20 million cubic metres. Currently, a further expansion of Rapid Gravity Filters by 90 mld has been commissioned. The responsibility of water supply to Chennai city is vested with Chennai Metropolitan Water Supply and Sewerage Board (CMWSSB).

Chennai Metropolitan Water Supply and Sewerage Board (CMWSSB)

The rapid increase in the growth of population in Chennai City warranted huge investment in water supply schemes. Since the available infrastructure was insufficient to cope with the requirements of an ever-growing population, it was thought that it would be desirable to form a separate Statutory Board to look after the water supply and sewerage system of Chennai City. The Chennai Metropolitan Water Supply and Sewage Board

(CMWSSB) was constituted in 1978, aimed exclusively for attending to the growing needs of water planning, development and appropriate regulation of water supply and sewerage services in Chennai City and Chennai Metropolitan Area. It gets institutional finance from national agencies like, Government of Tamil Nadu, Housing and Urban Development Corporation (HUDCO), Tamil Nadu Urban Financial Infrastructure Development Corporation (TUFIDCO), Life Insurance Corporation of India (LIC) and international agencies like the World Bank. The CMWSSB was constituted with the following objectives.

1. To promote or secure planned development, efficient operation, maintenance and effective regulation of water supply and sewerage system in Chennai Metropolitan Area (CMA).

2. To prepare long term plans to meet further requirements of water supply and sewerage in the CMA.

3. To undertake developmental activities with a view to meet the future requirements of water supply and sewerage in CMA.

4. To operate and maintain the water supply and sewerage system.

5. To discharge all such functions in relation to water supply and sewerage.

6. To undertake any other matter which is supplemental, incidental or consequential to all or any of the above functions, and

7. To undertake other functions as may be entrusted to the Board.

The progress of the activities after its establishment of CMWSSB is shown in Table 3. 5.

Table 3. 5. Progress made by CMWSSB, 1978 to 2008

Items	*1978*	*1998*	*1999*	*2000*	*2001*	*2002*	*2003*	*2004*	*2008*
Treatment Capacity (mld.)	182	620	620	627	627	627	627	627	1280
Water Supplied (in mld.)	180	440	485	195	200	180	300	660	645
Length of Water Main (km)	1250	1715	1853	2426	2330	2330	2330	2400	2924
Water Connection (Nos.)	116000	209000	228816	294338	294338	318780	326000	326094	499850
Distribution Stations (Nos.)	3	7	10	15	16	16	16	16	16

Source: Annual Report, 2003-04, CMWSSB. & & 2007-08 CMWSSB.

It is evident that the treatment capacities, length of water main and distribution stations have increased significantly between 1978 and 1998. During this period the quantity of water supplied increased from 180 mld to 440 mld, nevertheless an increase the number of water connections by two folds.

After 1998, though there was an increase in the number of water connections, distribution centres and length of water mains, the overall quantity of water supplied was less than the level recorded in 1998. Only two years witnessed an increase in water supply, the increase however, was only marginal. During the years 2000-2004, there was no change in treatment capacity of water supply, because poor rainfall was received in the catchments areas of water sources but is has increased to 1280 mld in the year 2008.

Institutional Arrangements in Water Supply

There are four main sources, which supply drinking water to the needs of the people coming under the purview of CMA. They are:

1) CMWSSB provides drinking water for the households in Chennai City.

2) It supplies water to municipalities and 9 towns adjoining Chennai City.

3) Some households and most of the industries make self-provision of water by drilling shallow wells or deep tube wells.

4) There exists private market in water supply.
   a. There is also bulk supply of water by means of tanker trucks of an average capacity of 12000 litres per day.
   b. There is retail distribution of bottled drinking water in plastic containers of 10 to 25 litres capacity.

Water Search

Due to a rapid increase in the demand for drinking water, the CMWSSB made several efforts to identify new sources of water. As a result of this search for additional water, CMWSSB, set up wells in Tamaraipakkam, Panjatty and Minjur fields, which resulted in the diversion of well water for domestic purposes. In the year 1983 a shallow aquifer in the coastal belt, south of Chennai was identified having a potential of about 100 mld of water. This aquifer is demarcated by the shallow stretch between Thiruvanmiyur and Muttukadu, covering a distance of about 18 Km. The water table in this area in summer is about 6 m below the ground level. This southern coastal aquifer was developed by a series of shallow tube wells and open wells and they have been functioning effectively.

Concerted efforts were taken in the year 1988 by CMWSSB. The ground water level in 1988 was around eight metres in the southern coastal aquifer. A main of 4-5 m was constructed in the 20 km stretch along the southern coast. In 1996, 96 licenses were issued for 260 firms for extraction and transportation of groundwater. During the 10-year period (1988-98), the average number of vehicles used per day for transportation of water annually raised from 15 in 1988 to 65 in 1998. A number of steps have been taken by the CMWSSB to conserve, manage and augment water resources in Chennai City.

Water Conservation

Rain Water Harvesting

Surface water is the major source of water for Chennai City, where no perennial rivers flow. During the North East Monsoon period, a huge amount of rainwater flow waste into the sea as floodwater. Therefore, Rain Water Harvesting (RWH) is used as one of the water conservation measures to recharge groundwater tables. In order to convert rainwater flow to potential use, CMWSSB initiated rainwater-harvesting (RWH) measures in 1994. As per the RWH regulation, new multi-storied or commercial complexes should incorporate RWH measures. Moreover, all central air-conditioned buildings have been asked to use treated sewage water for cooling purposes.

Further, CMWSSB has made it compulsory for the property-developers of multi-storied and special buildings to install suitable rainwater harvesting mechanisms, without which planning permission will not be accorded.

The CMWSSB has constituted for this purpose a fully dedicated RWH Cell. The main objective of the cell is to create awareness to the public on RWH and to offer voluntarily technical help on cost-effective methods of rainwater harvesting in their premises. A study was also carried out by CMWSSB, in Anna Nagar in Chennai City. Anna Nagar is one of the areas where RWH has been fully implemented by the residents. The details are given in Table 3.6.

Table 3. 6. Water Level in Anna Nagar, Chennai City

Year	*Water Level*	
	Rain Water Harvesting (mts)	*Non-Rain Water Harvesting (mts)*
1995-96	2.75	3.65
1996-97	2.50	3.80
1997-98	2.55	3.60
1998-99	4.00	4.50
1999-00	3.50	4.20

Source: Annual Report, 1999-2000, CMWSSB, p.23.

It is inferred that ground water was available within 2.75 meters during the year 1995-96 in the places of RWH whereas it was 3.65 meters in the Non RWH areas in the same year. It is understood that there is a significant difference in the ground water levels between RWH and Non RWH areas in Chennai City.

Control and Regulation of Extraction of Ground Water

Due to rapid depletion of groundwater, the water table has been falling in and around Chennai City. This leads to intrusion of seawater in some locations, thereby affecting the water quality. In order to control over exploitation of ground water, the Ground Water Regulation Act was enacted in the year 1987. Under the Act, the CMWSSB has been vested with the authority to regulate the extraction of ground water in the notified areas by a licensing mechanism.

Recharging of Ground Water

Ground water is an important source of water supply for Chennai City, next to surface water. Ground water level depends upon the rate of extraction. The CMWSSB heavily depends on ground water sources, consequent upon inadequate rainfall once in three or four years. Therefore, conservation of ground water is the important measure to guarantee water supply during the monsoon-failure years in Chennai City. In order to enable fast recharging of the aquifers during monsoon, three check dams were constructed in the year 1991 across the river Korataliyar.

These check dams have gradually raised the level of water table. In an effort to reduce the uneconomical and wasteful consumption of water, the system of recharging of ground water table has to be initiated.

In order to minimise the loss of water, the CMWSSB with the assistance of World Bank has undertaken certain re-engineering measures called *'leak detection'* and *'rectification programmes'*.

Leak Detection Porgramme

This control programme of the CMWSSB is aimed at managing the unaccounted water. The CMWSSB has carried out studies between 1989 and 1995 in two phases in an attempt to reduce leaks in the distribution system. The Phase-I study was conducted during 1989 to 1991. The outcome of the study was that all the house service connections did not have ferrule to ensure leak proof connection. Further, 44 per cent of these connections were leaking and 65 per cent of the conduit pipes are in bad condition.

The Phase – II studies have been taken up during the years 1994-95. They aimed at reducing the leakage level by providing ferrules, which were missing in the service connections. The major findings of the Phase – II study are that there prevailed 39 per cent leaks in ferrules, 31 per cent leaks in communication pipes and 39 per cent leaks in the mains. After rectification test was done, the achieved reduction in PVC main was 49 per cent and in CI 79 per cent. Based on the Phase II studies, the Phase III study was carried out to concentrate more on arresting leaks on a large scale. This phase covers 34800 house service connections in 40 corporation divisions. To remedy the situation limiting leakage to 151/c/hr., and renewal of 10 per cent of house communication pipes were undertaken. Tests were conducted for every 300 m of distribution main (Policy Note, Tamil Nadu Government, 2002-2003).

As a concomitance to Leak Detection and Rectification works, more house service connections were found leaking on the roadside and, therefore the test length of 300 metres was found to be too long. Plugging of the house service connections inside the property was also found difficult. Therefore the leakage limit was not attained. As a result of leak detection and rectification works, the leakage level was brought down from 30 per cent to 5 per cent, which is considerably a better achievement when compared to the internationally accepted leakage level of 10 per cent. Under Phase IV and V of leak detection and repair work, CMWSSB has taken up works in all the remaining 114 divisions. Under this work, 1354 km lengths of water distribution mains sizes up to 150 m have been tested for leaks. The work includes renewal of all the house service connections extending from the distribution main with non-corrosive flexible pipes up to 5 ft inside the property. The important achievements of leak detection and repair works include reduction in water loss, improved water supply, arresting water pollution and improvement in the quality of water.

Quality Control

During the period 1995-96, Quality Control System was introduced in order to standardise the quality of drinking water in Chennai City, which was later extended to water mains and sewers to meet consumer requirements. A separate Quality Control Wing is in existence exclusively to monitor the quality control aspects of the system. A separate mobile laboratory unit with two mini buses fitted with all necessary equipments is functioning for conducting on the spot study of the quality of water supplied to the consumers. To maintain the required level of residual chlorine even to the tail end of the water distribution system so as to curtail the water borne diseases such as cholera, automatic chlorinators have been built into the distribution system at critical locations. Initially, nine such automatic systems and another six chlorinators have been located in the head works or overhead tanks. To examine the status of water quality and quantity in the Southern Coastal Aquifer, Rail India Technical and Economic Survey conducted in the year 1996 an independent study on seawater intrusion in the southern coastal area.

WATER SUPPLY AUGMENTATION PROJECTS

First Chennai Water Supply and Sanitation Projects

The first Chennai Water Supply and Sanitation Project were taken up in 1988 at a cost of Rs.150.80 crores, aided by the World Bank. Under this project, the city water supply was augmented by a quantity of 50 mld per day by digging 30 bore-wells in 3 new fields, and 19 more bore-wells. A 300 mld capacity water treatment plant was set up to treat the first phase flow of

Krishna water. About 330 Kms of water mains have been rehabilitated and extended in the city and about 10 Km of existing trunk mains were strengthened (Annual report, 1995-96, CMWSSB, p.9).

The Second Chennai Water Supply and Sanitation Project

The Second Chennai Water Supply Project has been taken up for implementation in 1996 at a revised cost of Rs.778.79 crores, with World Bank loan assistance of US $.86.50 million. The funding pattern for the project includes 17.25 per cent from Government of Tamil Nadu as grant, and the World Bank as funds the balance of 65.5 per cent. The objective of the project is improvements in the sources of supply, distribution, conservation of water and expansion and rehabilitation of sewer collection, conveyance, treatment and disposal system. The expenditure on all these upto the year 2003 was Rs.639.35 crores (Annual Report, 2002-2003, CMWSSB, pp.9-10).

Chennai Water Supply Augmentation Project – I (New Veeranam Project)

The new Veeranam Project Works have been divided into 5 packages. The work commenced in 2002 and was completed in 2004. The total expenditure up to 2003 was 258.07 crores. The funding pattern of the project is given in Table 3.7.

Table 3.7. Funding Pattern of the New Veeranam Projects

Agencies	*Rs. in Crores*	*Percentage*
State Government contribution	239.36	33
Life Insurance Corporation of India (LIC)	182.19	25
TUFIDCO (Mega city)	100.00	14
TUFIDCO (Own fund)	50.00	7
Tax free bond	50.00	7
TNUIFSL	100.00	14
Total	721.55	100

Source: Annual Report, 2001-02, CMWSSB, p.23.

Third Chennai Project

The Third Chennai Project Proposals have been submitted at an estimated cost of Rs.600crores to the World Bank. The main aim of the Project is to draw ground water from Araniyar and Koratalliyar basin aquifer, augment water supply by constructing a check dam at Thirukandalam across Koratralliyar, refurbishment of Vallur anicut and desilting Rettai Eri at Madhavaram. It is also planned to extend the supply of water to the adjacent urban areas of Chennai City (Annual Report, 2001-02, CMWSSB, pp.9-10).

Chennai City River Conservation Project (CCRCP)

The Government of Tamil Nadu formulated the CCRCP at a cost of Rs.1700 crores with the aim of preventing sewage entering into the city waterways and also keeping them clean on a sustainable basis. CMWSSB shares Rs.720 crores of the project cost. The funding pattern is 68 per cent grant from National River Conservation Development (NRCD) and 32 per cent from other sources (Policy Note Government of Tamil Nadu, 2002-03).

Chembarambakkam Water Treatment Plant

The Master Plan contemplates increasing the water treatment capacity of about 830 mld owing to the increased supply of water from Telugu Ganga project and other local sources. Hence, an additional water supply treatment plant of 530 mld capacity was proposed to be constructed at Chembarambakkam. With the financial assistance of French Government, the project was undertaken by constructing intake tower, raw water pump house, treatment plant, water pump house and clear water transmission mains. The estimated cost of the project is about Rs.135 crore [Annual Report 2000-2001, CMWSSB, p.25]. Construction of the three zone reservoirs, one each at Valluvarkottam, Triplicane and Kannappar Thidal, has been taken up at a total project cost of Rs.25.55 crores.

Operation and Maintenance (O and M)

As far as O and M is concerned, it was a real paradigm shift from the concepts of *breakdown* maintenance to that of *preventive* maintenance. The single most important decision of the CMWSSB affecting the entire arena of O and M was to recognize and institutionalise preventive maintenance and to introduce necessary supporting systems. The CMWSSB created a separate Renewal and Replacement Fund to enable it to take care of financial requirements even in the years of deficit, without depending on the budgetary provisions. During the year 1998-99, a provision of Rs.10 crores was earmarked for this purpose (Annual Report 1998-99). A system of annual rate contracts both for the procurement of materials and standard works was introduced. Standardization of these procedures saved substantial costs and time, and increased the efficiency and productivity of the system. Successful Operation and Maintenance require quick internal communication facilities down the line. To achieve this end, field staff was provided with pagers, besides strengthening the wireless communication and mobile wireless facilities (Policy Note, Government of Tamil Nadu, 2003-2004).

Building Long Range Private Sector Partnership

In the fiscal year 1997-98, the CMWSSB allowed private sector to participate in Water Supply in order to raise the level of efficiency and cost effectiveness. The CMWSSB permitted private sector to operate in areas of establishment and operation, such as desalination plants and sewage treatment plants. This experiment encouraged the CMWSSB to further privatise its operations, because it tasted substantial gain in the maintenance and transportation expenditure. The CMWSSB saved RS.14.60 lakh in the operation and maintenance of wells in the fields. Again Rs.138.40 lakhs in the water transport through private operators and Rs.32.14 lakhs in maintenance of sewage pumping station in 1998 were spent by CMWSSB. The experiences encouraged the CMWSSB to transfer some of its public operations to private sector. The CMWSSB encourages demand-driven consumer oriented services.

Most of the water projects with private sector participation solely aim at an increase in production capacity. However, they do not address the main factors of inefficiency (Mehta, 1999). Harris (2003) doubts the types of rapid private sector participation in many inefficient and poorly performing utilities around the world. The improvements, according to him, under the existing framework tend to be unsustainable. The 2002 National Water Policy of Government of India encouraged for the first time private participation in planning, development and management of water resource projects for diverse uses, wherever feasible. Private sector participation is expected to produce innovative ideas, generate financial resources and introduce corporate management to improve service efficiency and accountability to users (Gol, 2002, p.5). Several recent studies have attempted to evaluate the effects of privatisation of water and other utilities on the welfare of the poor. Most studies focus on access to the water network and prices, since detailed information on water quality and regularity is not generally available.

In Chennai City, small-scale private operators sell bulk water through tanker lorries. The water sold is often untreated and unfit for human consumption. This market is completely exempted from regulation and control, viz., some of the companies are not registered and have no quality control mechanisms etc. They are mostly small operators employing a few people who were once owners of transport companies or of agricultural wells. This water business would experience a boom whenever, there is water crisis in Chennai City.

The CMWSSB, as reported by Ruet. et. al. (2002) provides tap water on an average of just 4 hours per day, and 97 per cent of the slum population of about 4 lakh households, are covered either by street water tap or tanker lories water services (MIDS

1995). The CMWSSB enjoys more autonomy than its counterparts of most other cities. It has contracted with 500 private contractors to supply water to various parts of the city, including slums that do not have public standpipes. Exnora, a Chennai-based Non-Governmental Organisation (NGO), works with the CMWSSB to organise the distribution of tanker waters to the slum dwellers. Some of the tanker transporters buy water from farmers outside the city by paying just Rs.3 per cu m and sell at Rs.15 per cu m to the CMWSSB (inclusive of cost of transport and maintenance). Approximately over 10 percent of the CMWSSB annual expenses go towards hiring and monitoring these tankers. Private tankers not under the contract to the CMWSSB buy water from the farmers and sell the water in bulk directly to the consumers at twice the price charged by CMWSSB. The CMWSSB, therefore, has a chance to monitor the quality of water and service provided by those under formal contract. However, the Board does not monitor the quality and service provided by the tanker lorries having no contract.

Financial Performance of CMWSSB

Expenditure

The CMWSSB has been maintaining the tempo of considering higher expenditures for infrastructure developments both in water supply and sewerage system. Table 3. 8 give the total expenditure of CMWSSB in the years 1994 to 2007-2008.

During the year 1994-95, the CMWSSB incurred an expenditure of Rs.75.02 crores on water supply. The expenditure increased to 242 crores in 2002-2003 and Rs.349.5 crores in 2003-2004, an increase of three times and five times respectively. Further it has increased to 488.6.0 in 2007-08. The major expenditure items are operation and maintenance and establishment expenditures, which are incurred for infrastructure development. It is interesting to note that debt-servicing expenditure occupied top position in almost all the years.

Table 3.8. Expenditure Items of CMWSSB from 1994-95 to 2007-08 (In crores)

Expenditure	*1994-95*	*1995-96*	*1996-97*	*1997-98*	*1998-99*	*1999-00*	*2000-01*
Operating and Maintenance	17.95	18.73	21.23	22.03	31.90	35.25	32.00
Payments and Provision to Employees	27.45	31.37	35.50	39.61	45.67	52.87	52.56
Office Administrative Expenses	1.35	1.44	1.86	2.39	3.86	3.93	3.26
Prior Period Adjustment	0.33	1.03	0.62	-	.74	2.03	6.07
Water Lorry Hire charges	4.34	6.53	5.66	6.17	7.69	9.20	23.86
Repair and Maintenance	9.09	-	-	-	-	-	-
Depreciation	6.09	11.57	13.22	14.59	17.20	27.04	38.88

(Contd.)

(Contd. Table 3.8)

Expenditure	*2001-02*	*2002-03*	*2003-04*	*2004-05*	*2005-06*	*2006-07*	*2007-08*
Operating and Maintenance	30.76	48.70	44.85	45.44	77.42	89.21	101.75
Payments and Provision to Employees	50.63	53.47	50.50	56.39	60.93	70.48	80.61
Office Administrative Expenses	2.51	2.86	2.95	3.85	4.55	4.65	4.08
Prior Period Adjustment	4.98	3.55	.0027	-	4.33	-	-
Water Lorry Hire charges	7.38	9.26	9.69	5.94	16.16	12.77	13.64
Repair and Maintenance	-	-	-	-	-	-	-
Depreciation	44.20		53.36	53.96	67.37	84.23	181.50

(Contd.)

(Contd. Table 3.8)

Expenditure	*1994-95*	*1995-96*	*1996-97*	*1997-98*	*1998-99*	*1999-00*	*2000-01*
Debt service charges	-	9.22	14.22	29.70	42.98	49.32	53.88
Provisions for Drought Debts: - Metered	-	-	1.58	.65	.03	-	-
Water charges – un metered	4.00	1.24		1.22		.23	.53
Interests on Loans	-	-	-	-	-	-	-
Taxes	4.42		0.04	2.18	1.27	1.60	2.49
Drought Expenditure							5.18
Total	75.02	81.13	93.93	118.54	151.34	181.47	218.77

(Contd.)

(Contd. Table 3.8)

Expenditure	2001-02	2002-03	2003-04	2004-05	2005-06	2006-07	2007-08
Debt service charges	56.15	66.84	78.43	-	85.89	87.62	95.86
Provisions for Drought Debts: - Metered	.81	1.03	-	-	-	-	-
Water charges – un metered	.69	.10	1.15	4.63	0.77	-	0.65
Interests on Loans	-	-	-	86.34	-	-	-
Taxes	1.16		.18	-	-	-	4.44
Drought Expenditure	82.69	8.01	108.39	224.46	7.88	11.10	06.02
Total	281.96	242	349.5	481.01	325.30	360.06	488.55

Source: Annual Reports, 1994-04, to 2007-08 CMWSSB.

Table 3.9. Income Sources of CMWSSB from 1995 to 2007-08 (Rs in Crores)

Income	*1994-95*	*1995-96*	*1996-97*	*1997 98*	*1998-99*	*1999-00*	*2000-01*
Sale of Water	66.58	73.98	84.41	101.97	122.92	136.19	137.05
Water Tax and Sewerage Tax	12.05	12.39	15.17	21.17	26.70	50.26	82.24
Grants and Subsidy from Government	-	-	1.00	1.0	-	-	-
Other Income	5.38	7.29	7.55	11.04	23.86	18.88	28.39
Excess Provision for Doubtful written back tax	-	0.24	0.33	-	-	-	-
Water Charges Method	-	1.73	-	-	0.10	1.66	0.35
Prior Period Adjustment	-	-	-	26.78	-	-	-
Devolution of Funds for water Drought	-	-	-	-	-	-	-
Taxes	-	-	-	-	-	-	-
Total	84.01	95.3	108.46	162.56	173.58	206.99	248.03

(Contd.)

(Contd. Table 3.9)

Income	*2001-02*	*2002-03*	*2003-04*	*2004-05*	*2005-06*	*2006-07*	*2007-08*
Sale of Water	126.64	146.67	169.4	189.15	205.33	200.29	204.04
Water Tax and Sewerage Tax	56.78	57.55	57.60	59.36	66.66	65.88	83.07
Grants and Subsidy from Government	75.00	10.44	09.07	203.88	12.30	24.76	9.10
Other Income	25.07	29.18	23.37	0.27	28.40	29.89	44.93
Excess Provision for Doubtful written back tax	-	1.31	.69	-	-	-	1.28
Water Charges Method	-	-	-	0.48	0.37	1.93	-
Prior Period Adjustment	-	-	-	27.24	-	2.46	112.47
Devolution of Funds for water Drought	-	5.00	92.00	-	-	-	-
Taxes	-	-	-	0.60	0.10	-	-
Total	283.49	250.15	352.15	481.01	325.03	360.06	488.55

Source: Annual Reports, 1994-04, CMWSSB.

Income

Table 3.9 indicates the income of CMWSSB. During the year 1994-95, the total income of the CMWSSB was Rs. 84.01 crores, it increased to Rs. 250.15 crores in 2002-2003 and Rs, 352.13 crores in 2003-2004 and 45 crores in 2007-08. The CMWSSB continued to record increasing income over the reference period of time.

Trends on Income and Expenditures of CMWSSB

Table 3.10 indicates the trends in income and expenditure of CMWSSB from 1994-95 to 2003-2004.

Table 3.10. Trends in Income and Expenditure of CMWSSB from 1994 -95 to 2007-08

(Rs in Crores)

Years	*Income*	*Expenditure*	*Differences*
1994 - 95	84.01	75.02	8.99
1995 - 96	95.63	81.13	14.50
1996 - 97	108.46	93.93	14.53
1997 - 98	162.56	118.54	44.02
1998 - 99	173.58	151.34	22.24
1999 - 00	206.99	181.47	25.52
2000 - 01	248.03	218.77	29.26
2001 - 02	283.49	281.96	1.53
2002 -03	250.15	242.00	7.58
2003 - 04	351.45	349.50	1.95
2004 - 05	481.60	476.42	65.18
2005 - 06	313.20	325.74	-12.54
2006 - 07	324.12	360.09	-35.97
2007 - 08	454.91	488.60	-33.69

Source: Annual Reports, 1994-04, CMWSSB.

During the last 10 years, the CMWSSB continued to record surplus income. During the year 1997-98, the net surplus amount was Rs.44.02 crores. The increase has been possible mainly due to an upward revision of water charges. The income recorded for the year 1998-99 and 1999-2000 were Rs. 22.24 crores and 25.52 crores respectively. But compared to the year 2002-2003, the net income of the CMWSSB has been on the decline in the year 2003-04. Further it has declined to -33.69 crores in 2007 – 08. It is noted that the expenditures exceed over the income.

Financial Reforms

a) Implementation of Tariff Policy

A sound tariff policy is the backbone of any viable financial management system. In order to improve its financial position, a new tariff policy was adopted by the CMWSSB in 1997. The major initiatives were measures taken to further improve and strengthen some of the financial systems. They include i) rationalization of tariffs to end-use consumers with effect from 1998, reclassifying them with reference to the end-use of water supplied. ii) rationalisation of consumers and giving relief to small commercial establishments, attached to domestic properties, iii) revision of tariff. The details of implementation of the new tariff policy are shown Table 3.11.

Urban water supply is a state subject. Maintaining capital assets and fixing and collecting water tariffs are the responsibility of the respective local bodies. But most of the urban local bodies are found to be weak financially. The urban local bodies mobilise revenue from internal and external sources. Internal sources include water tax, drainage tax, etc., and non-tax sources include water charges, drainage charges etc. The external sources include loans and grants from the state government and international financial institutions. However, the percentage of plan outlay for rural water supply has been more than that in urban areas after the Fifth Five Year Plan.

High municipal water connection fees restrict the households in lower income groups from seeking water

Table 3. 11. Chennai Metropolitan Water Supplies and Sewerage Board Tariff Structure per Month

Sector	*Slab*	*Tariff Rate in Rs / KL*						
		1986	*1987*	*1992*	*1994*	*1997*	*1998*	*2008*
Domestic (metered)	Up to 10 KL	-	-	-	-	1.30	2.50	2.50
	11 KL to 15 Kl	-	-	-	-	-	10.00	10.00
	15KL to 25 KL	-	-	-	-	13.00	15.00	15.00
	26KL to 50 KL	-	-	1.00	-	39.00	-	-
	Above 50 KL	0.50	1.00	2.00	30=pm	-	25.00	25.00
Unmeterd	-	-	2.00	12.00/pm	30.00/pm	40.00/pm	50/pm	50
Commercial metered	Up to 50 KL	-	12.00/pm	3.00	-	-	25.00/pm	35.00
	Above 50 Kl	1.00	3.00	4.00	10.00	15.00	40.00/pm	60.00
Unmeterd	-	-	4.00	60.00/pm	125.00/pm	165.00/pm	400.00/pm	650.00

(*Contd.*)

(Contd. Table 3.11)

Sector	*Slab*	*Tariff Rate in Rs / KL*						
		1986	*1987*	*1992*	*1994*	*1997*	*1998*	*2008*
Institutiona (Metered) Unmetered	Total Quantity	2.00	60.00/pm	2.00	10.00	15.00	20.00/pm	30.00
Bulk Supply		15.00/pm	2.00 60.00/pm	60.00 75.00/pm	125.00/pm 250/pm	165.00/pm 330.00	200.00/pm 600/plocal	300.00 600.00
		-	-	7.00	25.00	35.00	-	-
Industry metered	Permitted supply or upto 50 kl	1.50	3.00	10.00	125.00	35.00	-	-
un metered	Beyond	2.00	4.00	75.00/pm	250.00	330.00	-	-
	permitted supply	-	60.00/pm	75.00/pm	-	-	-	-

Source: Annual Reports, 1986-08, CMWSSB.

connection. Even though Chennai City has simplified these procedures, it is usually very difficult to get new connections. The cost of getting new water connection in Chennai city is Rs.2000. The poor households can afford to pay monthly charges, but they are unable to arrange the entire cost of new connection in one instalment. Fixation and revision of water charges are linked to a strong political decision (Moench and Janakarajan, 2002). The usual procedure is a proposal for tariff increase proposed by the water boards and this needs to be approved by the elected councillors. It might take years for tariff increase to be approved due to populism and strong vote bank politics. Due to low water tariff rate, no policy can be sustainable in the long run.

It is evident that the measures undertaken by the government to augment water supply situation in Chennai is laudable despite some limitations in the demand and supply management. The marginal cost of water supply is very high due to the tapping of water from distant sources and the absence of a perennial water augmentation source.

Reference

1. Chennai is the new name of Madras City. The Madras City was renamed as Chennai in 1996 by an Act of 1996, called the Alternative of Name Act.

4

Water Supply Management in Chennai City

Every year, Chennai City is facing water scarcity. Frequent water supply crisis and unreliable water supply are the common phenomena of Chennai City. The increasing demand for domestic water supply, for an ever-growing population, the rapid growth of urbanization and unplanned measures put enormous stress on the limited available but limited water resources. With the result, the per capita water supply in Chennai City is the lowest in 2001 (76 lpcd), when compared to the major metropolitan cities of India (Joel Ruet, Saravanan and Maire-Helene Zerah, 2002). Since the quantity of water supply has been dwindling, the water supply to households in Chennai City is often restricted to every alternative day. It further deteriorates in the drought years (2001-2003). Availability of water has become a politically and environmentally sensitive issue (Anil Agarwal, 1997). The main reason for water scarcity is the absence of perennial sources of water and therefore it has to depend on a few freshwater reservoirs. Hence, water supply management is an important measure to provide water to the requirements of households in different parts of Chennai City.

The present chapter focuses mainly on water resources management, especially on the extent of water supply and demand side management. On the supply-side management, this chapter discusses the supply of water from various surface water sources and quantity of surface and ground water production. It also concentrates on drinking water requirements, household water supply and water crisis. It also analyses the non-conventional alternative water sources that are available to fill the gap between demand and supply.

The primary sources of water supply to Chennai City are surface water and ground water. Surface water is the basic water source, where the run-off water during the monsoon is stored in tanks and reservoirs. The stored water is then treated by the Chennai Metro Water Supply and Sewerage Board (CMWSSB) and subsequently supplied to different distribution centres. Ground water tapped by individual households from dug-wells and bore-wells also supplements the surface water source.

The responsibility of water supply in Chennai City rests with the CMWSSB. It prepares master plans, schemes of operation and management. It secures water from various sources during the period of water crisis. The Board forecasts demand and supply of water for both short-run and long run. The CMWSSB implements various water supply schemes in order to augment the quantity of water supply to various needs of people, industry and institutions in Chennai City. As per the Master Plan for the Horizon Year 2011, the CMWSSB, has planned to increase the water supply from the current level of 70 lpcd to 121 lpcd in 2011.

Sources of Water Supply

Surface water

The main sources of surface water to Chennai City are the three reservoirs viz., Red Hills, Cholavaram and Poondi. Chembarambakkam tank has been functioning as an additional source of surface water from January 2000 onwards, besides Erattai Eri and Porur Lake. The storage of water in these reservoirs purely depends upon the availability of rainfall in catchments areas. During rainy seasons, these lakes are filled with run off of the rainwater. The latest addition to the Chennai water supply system is the New Veeranam Scheme, which serves as an important supplementary source.

The total capacity of the surface water sources for Chennai City is shown in Table 4.1.

Table 4.1. Storage Capacity of Surface Water Sources

Water Supply Sources	*Capacity (in Mcft)*
Poondi	3231
Cholavaram	881
Redhills	3300
Chembarambakkam	3645
Rettai Eri	32
Porur	15
The New Veeranam	1485

Source: CMWSSB, Annual Report, 2004.

Increase in the feeder-flow to reservoirs for meeting the ever-rising drinking water demand of Chennai City was taken up with the diversion of floodwaters of Koratalaiyar to the Redhills Lake, in the later part of the 19th century. The man-made reservoir at Poondi was constructed in 1948. Major rainwater collection storages of Poondi, Cholavaram and Redhills are shallow, where evaporation has a decisive effect on the availability of storage water for supply. The total annual evaporation from all the three reservoirs was estimated on an average at 75 million cubic metres from 1950 to 1983.

Ground Water Sources

Owing to rapid population growth and poor storage in reservoirs as a result of inconsistent rainfall, the availability of surface water is insufficient to meet the drinking water demands of Chennai City. In order to get additional sources of water, attention is being shifted to alternative sources of surface water. In this effort, the area North – East of Chennai City was taken up for extensive hydrogelogical study in the year 1965 by UNDP to assess the potentiality of ground water source (Master Plan for CMWSSB-2011, 1991). Based on the information supplied by the study, CMWSSB developed well fields at Poondi, Tamaraipakkam, Flood Plains, Kannigaiper, Panjety, and Minjur. Later the Southern Coastal Aquifer was identified in the area and over 74 deep bore wells are in existence in the area. The total supply of ground water from all these wells accounted for 111.5mld in 1999-2000. Moreover the ground water supply from the seven wells fields in the North – East of Chennai City and the Southern Aquifer was 144mld in the same period. The yield of groundwater sources in 1999-2000 is shown in Table 4.2.

Table 4.2. Ground Water Supply, 1999 – 2000

Ground Water Source	*Designed yield (mld)*
Tamaraipakkam	29.25
Panjetty	36.00
Minjur	24.75
Poondi	27.00
Floodplains	13.50
Kannigaiper	13.50
South Chennai Aquifer	6.80
Total	150.80

Source: CMWSSB Annual Report, 1999-2000.

In addition to the extraction of ground water by the CMWSSB for public distribution, individual households and private water suppliers also extract ground water, especially for non-drinking and cooking needs. Due to the indiscriminate exploitation of the ground water, the water level has declined in many parts of Chennai City. The zone-wise average ground water levels in Chennai City are furnished in Table 4.3.

Table4.3. Zone wise Average Groundwater Level in Chennai City in 2004

Zone Number	*Zone Area*	*Ground water level (in metres)*
I	Erukkanjeri, Tondiarpet, Washermanpet	5.14
II	Royapuram, George town, Beach station	7.09
III	Vepery	4.64
IV	Villivakkam, Sembiyam, Perambur, Aynavaram	7.50
V	Anna Nagar, Chetpet	6.85
VI	Park town, Triplicane	5.15
VII	Purasawakkam, Egmore, Royapettah	5.65
VIII	Kodambakkam, T.Nagar	7.71
IX	Saidapet, Kotturpuram	7.40
X	Mylapore, Adyar, Thiruvanmiyur	6.56

Source: The Hindu, Deficit rainfall tells on groundwater table, 7th February 2004.

Waterways in Chennai City

Chennai City is traversed by six waterways namely Cooum River, Adyar River, Buckingham Canal, Otteri Nullah, Captain Cotton Canal and Mambalam Drain. The waterways of Chennai are not perennial; nevertheless they function as sewerage conduits.

River Cooum is a surplus drain of a number of tanks having a length of 65km and it passes through Chennai City limit by 17.98 kms. Adyar River is also a short one of about 42 kms. It traverses a distance of 14.6kms in the city through Nesapakkam, Saidapet, and Nandanam. The average annual flow of Adyar has been computed as 140mcm. However, it carries floodwater during monsoon causing inundation and damages along its banks. Below Maraimalai Adigalar Bridge at Saidapet, the river is marked by the tidal inflow due to seawater. The available water of this basin has not been tapped for meeting the city's water needs, because drainage water is diverted into this river making the water unuseful. Buckingham canal also receives significant quantity of sewage and floodwater from both east and west sides of the city. No water is drawn from this canal for any kind of consumption purpose. The Captain Cotton Canal (CCC) is the smallest waterway running through parts of the city north of the Cooum River. The canal courses for a length of 2.9 km through the inhabited areas of Erukanchery, Vyasarpadi and Korukkupet. Otteri Nullah is also a small waterway traversing through Chennai City north of the Cooum River, originating at Anna Nagar and joining the North Buckinham Canal near Basin Bridge after traversing a length of 8.7km. The Mambalam drain originates from the central part of the city, near Valluvarkottam and after traversing through T.Nagar, Mambalam, CIT Nagar, Saidapet, Ashok Nagar, Jafferkhanpet, and Nandanam it drains into Adyar River (CPCB, 2001, pp.13-18).

None of the above waterways of Chennai City is used presently for drinking water purposes because the waterways function as dumping grounds and carriers of domestic, commercial and industrial wastes and effluents. The uncollected sewage from unorganised sector, sewage from treatment plants, effluent discharged from commercial establishments and waste water from drainage systems finally reaches the waterways of Chennai City.

Water Production

Surface Water Production

Surface water stored in reservoirs for meeting the water demands of Chennai city have often proved inadequate, especially when there is a shortfall of rain in and around the catchments areas of these reservoirs. The shortage of rainfall not only

Table 4.4. Combined Storage of Water Reservoirs for Chennai City, 1994-2004

Year	*Rainfall in city Reservoirs (Normal Rainfal-1400, mm)*	*Average storage of Poondi Reservoirs (Total Capacity, 3231, mcft)*	*Average Storage of Cholavaram Reservoirs (Total Capacity, 881 mcft)*	*Average Storage of Redhills Reservoirs (Total Capacity 3300,mcft)*	*Average Combined Storage (Total Capacity, 7412, mcft)*
1994	1282	1272	399	2267	3938
1995	1566	1299	444	2666	4408
1996	2123	1251	557	2335	4408
1997	2169	1775	326	2292	4393
1998	1334	1678	454	1975	4107
1999	624	1064	260	2252	3576
2000	1053	115	43	1111	1269
2001	1400	300	11	725	1035
2002	1129	740	102	2023	2865
2003	810	115	66	1076	1257
2004	744	36	15	186	237

Source: CMWSSB, http:/www.chennaimetrowater.com/lakelevel/. Html, 2004.

reduces the storage of surface water to a larger extent but also brings down the ground water level. The availability of combined storage water in the reservoirs over the years is furnished in Table 4.4.

The normal maximum combined storage capacity of Poondi, Cholavaram and Red Hills is 7412mcft. The maximum storage level of 4408mcft was recorded in 1995 and 1996, which was around 35 per cent of the total storage capacity. In 2004 the combined storage of the reservoirs reduced to the lowest level of 237 mcft, which was the lowest minimum recorded in the last 35 years. The previous lower level was 880 mcft recorded in 1987.

It is due to the fact that the average rainfall recorded in 2004 was 744 mm against the normal annual rainfall of 1400mm. From this it is clear that the storage in all the three water reservoirs in Chennai City totally depends upon the rainfall in respective catchments areas. Even a single failure of monsoon in one particular season influences the storage level in the reservoirs.

Ground Water Production

The annual average availability of water from the well fields for providing water to Chennai City is shown in Table 4.5.

Table 4.5. Well Field Production

Year	*Average Well field Production (MLD)*
1991	69.27
1992	66.01
1993	52.24
1994	51.67
1995	54.85
1996	52.02
1997	53.68
1998	39.08

(Contd.)

(Contd. Table 4.5)

Year	*Average Well field Production (MLD)*
1999	39.61
2000	79.48
2001	86.65
2002	72.97
2003	90.16
2004	85.75

Source: CMWSSB, 1991-2004.

After the year 2000, domestic water supply to Chennai City has been largely supported by the production of water in the well fields. The well water is normally supplied through tanker lorries. The dependence on ground water to meet the needs of water to Chennai City increased in the first quinquinnium of 2005. Except in the good monsoon years of 1998 and 1999, the level of exploitation of groundwater has been on the increase.

Water Requirements

The requirement of water for Chennai City was estimated by CMWSSB in the year 1991, on the basis of a supply of 150 lpcd(litres per capita consumption per day) for Chennai City and 100 lpcd for other towns in CMA. Multiplying population with 150 lpcd estimates the total requirement of water supply. Overall requirements are inclusive of domestic, commercial,

Table 4.6. Water Requirements in Chennai City

Use category	*Unit water requirements (mld)*		
	1981	*1991*	*2001*
Domestic purpose (including commercial, institutional and slums)	492	576	631
Industrial purposes	150	160	200
Total requirement	642	736	831

Source: CMWSSB, 1991.

institutional, industrial and slum water requirements. The details of domestic and industrial water requirements of Chennai City are shown in Table 4.6.

The overall requirements of water per day for Chennai City in 1981 were estimated as 642 mld which increased to 831 mld in 2001 due to increase in the population and economic activities.

Water Demand for Industrial Purposes

Mostly, groundwater is used for industrial purposes. During the year 2002-2003, a number of industries suffered, because groundwater was used for drinking purposes. Water-intensive industries like the Chennai Petroleum Corporation Limited and the Madras Fertilisers Limited had to shut down their operations for some days due to shortage of water. As industry sources point out, a number of fly-by-night water tanker operators got into the business and they not only pushed down the ground water levels but also affected the water quality. With water scarcity expected to be a perennial problem, industries have adopted different methods to deal with the water crisis. For example, Ashok Leyland Limited, which has a plant making commercial vehicles at Ennore, requires about 12 lakh litres of water a day for all purposes is depending on the 115 open wells in its campus (CMDA, Chennai, 2001, 1991).

The Chennai Petroleum Corporation Limited (CPCL) requires about 18 mld of water. Nearly, seven mld for producing demineralised water for generating steam and 22 mld for cooling tower and drinking water requirements are met from CMWSSB. The CMWSSB supplies the CPCL 11 mld of raw water, of which nine mld from secondary treated sewage and 1.5 mld from effluent water. According to the CPCL, it spends about Rs.15.5 crore for raw water and Rs.2.4 crore for secondary treated sewage water every year. CPCL requires an additional 15 mld of water for its expansion project. According to official sources, the

CMWSSB supplies 50 mld of water for industrial purposes in Chennai City (Industrial Potential Survey Report, Chennai (2001-2002).

Development of Water Distribution System in Chennai City

The surface water from all the reservoirs of Chennai City reaches Red Hills and then it is transmitted through a closed conduit to Kilpauk transplant plant and pump house. Chennai water distribution begins after treatment from Kilpauk. The treated water is pumped to various distribution stations. The water from well fields is directly connected to the main, carrying water to industries in the north.

In order to remove the existing systemic deficiencies and to ensure equitable distribution of water throughout Chennai City, CMWSSB has executed a water supply and sewerage

Table 4.7. Capacity of Zonal Head Works Zones / Station

Station	*Storage in UGT*	*Capacity in Cu.M. OHT*	*Supply Mld*
Choolaimedu	43.00	5.00	30
Ekkattuthangal	4.40	0.60	7
K.K. Nagar	14.00	2.40	30
Kulathur	20.94	-	15
Mylapore	24.50	-	15
Nandanam	11.52	1.20	5
Pallipattu	17.00	2.40	15
Patel Nagar	13.00	-	14
Velachery	22.00	-	7
Vyasarpadi	24.80	-	15
Anna Poonga	13.99	2.60	20
Kilpauk	35.51	9.00	60
Madhavaram	9.68	-	40
Valluvarkottam	15.00	2.90	20
Triplicane	10.00	2.40	5
Kannapar Thidal	19.00	3.00	10

Source: CMWSSB, 2003-2004.

master plan. It contemplates introducing the concept of self-sustaining zonal distribution system. Each zonal distribution system has a reservoir, distribution station and necessary pumping system. The net-works of water distribution arrangements divide the entire city into different zones, each with a separate headwork. The required infrastructure for the zonal distribution system was also developed. The capacity of zonal head works is given in Table 4.7.

The present water supply system in Chennai City functions with 5-distribution zones, viz., Kilpauk, Southern Head Works, Anna Poonga, K.K. Nagar and Pallipattu. The water supply and sewerage master plan of CMWSSB has been organised in such a way that it would ensure a minimum residual pressure of at least 7 metres at the ferrule point of each and every house service connection pipe. The zonal distribution systems have been designed in such way to meet the water demand of Chennai City, 10 Adjacent Urbanised areas and Manali industries.

Quantity of Water Supply in Chennai City

The important inference is that the average monthly supply of water over the past 35 years did not show much progress. Nevertheless, the variations in water supply in different months in a year purely depend on the variations in the rainfall. Generally from September to December, the monthly average supply of water seems to be better than the annual average. Table 3.8 gives the details of monthly average water supply to Chennai City by the CMWSSB from 1971 to 2004.

Table 4. 8 reveals that the average supply of water in mld is not as per the demand for water in Chennai City but it is based on the availability of water. Moreover, the variation ranges from the lowest of 140 mld in 2003 to the highest of 410 mld in 1999.

Table 4.8. Quantity of water supplied in Chennai City (mld) Per Month, 1975-2004

Year	*Jan*	*Feb*	*Mar*	*Apr*	*May*	*Jun*	*Jul*	*Aug*	*Sep*	*Oct*	*Nov*	*Dec*	*Average*
1975	135	103	90	95	73	76	73	78	146	156	175	179	115
1976	185	191	196	201	203	203	205	203	202	198	185	204	198
1977	209	214	215	212	226	228	229	231	231	231	229	225	223
1978	226	229	232	235	244	249	249	249	246	246	232	239	240
1979	240	237	240	239	236	244	245	246	245	247	250	248	243
1980	246	242	243	246	253	250	246	235	237	242	237	236	243
1981	235	234	237	237	220	214	202	184	179	179	249	248	218
1982	250	245	247	245	241	229	211	196	175	156	185	179	213
1983	169	130	93	98	105	102	100	110	124	197	231	222	140
1984	210	202	225	243	241	244	235	239	245	252	237	250	235
1985	246	254	246	250	243	236	232	231	227	230	232	243	239
1986	263	271	258	261	260	259	259	253	243	248	257	250	257
1987	229	193	176	169	170	124	101	107	103	102	121	145	144
1988	128	134	137	155	149	152	162	146	159	183	176	168	154
1989	169	141	158	162	150	149	161	149	135	146	135	139	149

(Contd.)

(Contd. Table 4.8)

Year	*Jan*	*Feb*	*Mar*	*Apr*	*May*	*Jun*	*Jul*	*Aug*	*Sep*	*Oct*	*Nov*	*Dec*	*Average*
1990	145	136	152	143	139	136	134	139	137	140	228	228	153
1991	229	230	233	229	229	227	228	228	228	228	230	251	231
1992	251	252	252	252	253	254	254	253	253	254	254	252	253
1993	175	119	116	117	104	102	100	86	85	85	101	155	112
1994	159	168	171	252	255	258	255	257	258	256	262	277	237
1995	279	274	225	276	281	286	287	302	330	332	329	317	297
1996	249	250	248	251	251	251	274	276	282	411	411	399	296
1997	434	412	363	360	382	357	360	313	265	251	277	408	348
1998	417	425	394	364	345	334	339	346	365	374	422	437	380
1999	446	423	443	450	455	458	400	452	436	387	339	232	410
2000	214	209	207	209	205	199	186	184	194	218	200	182	200
2001	184	180	294	207	212	139	115	114	113	116	139	147	155
2002	208	214	235	326	255	357	353	340	318	291	263	250	292
2003	192	150	154	150	141	138	128	222	125	134	125	123	140
2004	128	125	126	130	138	146	151	155	157	206	261	274	167

Source: CMWSSB, Quantity of water supplied to the City (in Million Litres), 1975-2004.

Table 4. 9. Quantity of Water Annually Supplied to Households and Industries in Chennai City, 1975-2004

Total Water supplied to the city in mld (a)	*Average Quantity of water supplied to the city in mld (b)*	*Water supplied to the Industry in mld (c)*	*Average Quantity of water supplied to the Industry in mld (d)*	*Total (a+c)*	*Total average quantity of water supplied to city in mld (b+d)*
1368	114	-	-	1368	114
2376	198	-	-	2376	198
2676	223	-	-	2676	223
2868	239	-	-	2868	239
2916	243	-	-	2916	243
2904	242	-	-	2904	242
2604	217	343	28	2947	245
2556	213	552	46	3108	259
1680	140	336	28	2016	168
2820	235	504	42	3324	277
2856	238	576	48	3432	286
3072	256	612	51	3684	307
1728	144	444	37	2172	181
1848	154	564	47	2412	201

(Contd.)

(Contd. Table 4.9)

Total Water supplied to the city in mld (a)	*Average Quantity of water supplied to the city in mld* (b)	*Water supplied to the Industry in mld* (c)	*Average Quantity of water supplied to the Industry in mld* (d)	*Total* (a+c)	*Total average quantity of water supplied to city in mld* (b+d)
1788	149	504	42	2292	191
1836	153	528	44	2364	197
2760	230	564	47	3324	277
3024	252	600	50	3624	302
1344	112	384	32	1728	144
2820	137	444	37	3264	174
3564	197	468	39	4032	236
3552	196	456	38	4008	234
4176	348	492	41	4668	389
4560	380	492	41	5052	421
4920	410	480	40	5400	450
2400	200	576	48	2976	248
1860	155	444	51	2304	206
3504	292	552	46	4056	338
1680	140	588	49	2268	189
1998	166	602	50	2600	216

Source: CMWSSB, Annual Reports, Quantity of Water Supplied to the City and Industries Per Year, 1975-2004

Table 4.10. Decadal Population Growth, Water Demand and Supply in Chennai City, 1981-2001

Year	*Population in million*	*Water demand in (mld)*			*Average Quantity of Water Supplied (mld)*			*lpcd*
		Households	*Industries*	*Total*	*Households*	*Industries*	*Total*	
1981	3.28	492	150	642	217	28	445	66
1991	3.84	576	160	736	230	47	277	59
2001	4.21	631	200	831	155	48	203	37

Source: S. Nambi Ayyadurai 1993, "Reuse of wastewater" Master plan of CMWSSB, and International Symposium on Strides in Civil Engineering.

Total quantity of water annually supplied to the households and industries in Chennai City is shown in Table 4.9.

It is clear that fluctuations in the quantity of water supplied to households are more when compared to a relatively stable supply of water to industries over the years.

Short fall of Water Supply

One of the most important causes of water crisis in Chennai City is the impact of demographic changes. While reviewing Table 3.9 it is clear that absolute water balance is on the decline in per capita availability, owing to an almost fixed supply of water to an increasing population. The expansion of industrial activities without a proper plan on water management is also an important reason for the problem of water stress. The details of average supply and demand for water inclusive of domestic and industrial needs are furnished in Table 4.10.

Table 4.10 indicates the increased water demand for domestic and industrial uses. In 1981, total domestic and industrial water demand was 642 mld, which increased to 831 mld in 2001. However, the average supply to the domestic use recorded a moderate increase in 1991. Still, it drastically declined in 2001. The intensity of water crisis in the three decades prior to 2001 can be better understood by taking note of the decline in lpcd.

Water Supply Scenario of Chennai City, 2001-2003

Chennai City witnessed unprecedented water crisis in 2001. The average quantity of water supplied was 155 mld which covered only 65 per cent of the population, at the rate of 59 lpcd. The rest of the people were supplied water through non-piped distribution system. There were 4524 water storing tanks installed in different parts of the city and altogether 0.44 million liters were supplied through tanker lorries. For washing and other purposes, the CMWSSB installed 4166 hand pumps. To maintain Chennai City water supply at the existing level, the CMWSSB had taken some contingency measures to transport water from distant sources, such as Neyveli, Mamundur, Palur and Kannigaiper by road and even from Mettur and Erode by train (CMWSSB, Annual Report 2000 – 2001).

During the year 2002, the North East monsoon was normal in Chennai City. But in the catchment areas of reservoirs supplying water to Chennai City, the rainfall was deficient. The total water supplied to the city was 3504 mld in the year 2002. The average quantity of water supplied was 292 mld. In order to maintain regular water supply, about 3.25 tmcft water was drawn from Krishna River by the CMWSSB in the year 2003.

The years 1999 and 2003 were the worst in terms of rainfall received by the combined storage reservoirs. According to CMWSSB statistics, the rainfall registered in the catchments area of the reservoirs in the year 2003 was the second lowest in the last 10 years. The Northeast monsoon helped to increase the storage points by 385 mm, when compared to the average of 753 mm. The combined storage of the city reservoirs was 291mcft against its total capacity of 7412 mcft. In order to ensure supply of drinking water to the residents in this difficult situation, the CMWSSB has implemented the programme of supplying of water on alternate days through the distribution pipelines. Alternate day's supply is so maintained as to ensure that water reaches the tail-end areas. However, tanker lorry water supply was augmented to maintain water supply to defective and un-served streets to the extent required. About 275 mld was supplied on supply days through pipeline and 25 mld on non-supply days by means of lorries for maintaining water supply. The drawal from Red Hills Lake was also supplemented by ground water extraction from well fields to an extent of about 80 mld, which gradually increased to 100 mld. The CMWSSB also hired additional 75 numbers of private agricultural wells to augment the yield about 25 mld of water to stabilise the water supply during summer. Infrastructure improvements in the well fields was taken up at a cost of Rs.12.58 crores in 2002-2003 for providing additional pipelines and pump sets to carry through the additional water from agricultural wells. A Contingency Plan Year had been drawn up by the CMWSSB at a cost of Rs.55 crores to execute the infrastructure facilities required to supplement the available water from the reservoirs.

Status of Chennai City Water Supply in 2004

Rainfall due to South West monsoon during 2003 was below normal and the North East monsoon was a total failure in Chennai City. As a result, the storage in the reservoirs on the first month of 2004 was 272 mcft, which was about 3.7 per cent of the total storage capacity. Further, the storage level declined to 198 mcft in April 2004 and to 134.50 mcft in June 2004. However, regulated supply of water on alternate days was maintained and precautionary steps were taken to maintain supplies without causing undue distress to the people. Even though there was inadequate water storage in the reservoirs of Andhra Pradesh, water was pumped from dead storage of Kandaleru and 207.10 mcft of water was received at zero point in Poondi Reservoir in Tamil Nadu from March to April 2004. After that, the pumping operations could not be continued. The water supply composition in 2004 was entirely different from the water production under normal conditions. Table 4.11 gives water production under normal conditions and Table 4.12 gives the water supply position in the drought year of 2004.

At present, water supply to Chennai City is being maintained only by drawals from the well fields and also by transporting water to an extent of about 90 mld per day in tankers from adjacent areas located within a radius of about 100 Kms. The quantity had been enhanced in a phased manner and at present more than 85 mld of water per day is transported through hired mobile tankers by CMWSSB. By the end of June 2004, about 2663 tanker lorries had been pressed into service.

Table 4.11. Water Production under Normal Conditions

Source	*Location*	*Water Obtained (MLD)*
Surface	Poondi, Redhills, Cholavaram, ,	200
	Veeranam Lake	180
	Kandaleru dam	130
Ground Water	Wells in A K, Southern Coastal	233
	Bore Wells in Nayveli	90
Total		833

Source: CMWSSB – Annual Report, 2002-2003.

Table 4.12.Water Production during the 2004 Drought

Source	*Location*	*Water Obtained (MLD)*	
		June -04	*From October -04 to January-05*
Ground Water	Transported through tanker	98	30
	Metro wells along the northwest	95	450
	Southern Coastal Aquifer	3	-
	From Palar river bed	2.5	-
Surface	Redhills	2	40
	Desalination Plants	0.5	-
	Veeranam lake	-	180
Total		201	700

Source: CMWSSB Annual Report, 2003-2004.

This number has gradually declined after the new Veeranam scheme was made available to the city in 2004. Besides this, using 1219 lorries of each having 9 KL or 6 KL capacity, totalling 11400 trips per day, internal distribution was carried out (Industrial Economists, 2004, pp.18-22).

Impact of Water Crisis in Chennai City

Rapid increase in population and dry weather conditions combine to produce large-scale water shortages. As human population grows and standard of their living improves, there is an increase in the demand for fresh water. The main objective of the International Drinking Water Supply and Sanitation (IDWSS, 1977) was to provide safe drinking water and adequate sanitation to the entire population by the year 1990. But two billion people over 40 nations face the problem of freshwater shortage. The per capita water consumption in industrialised countries (500-800litres per day) is on an average 10 times more than that of developing countries (60-150lpcd). The combined effects of urbanisation, industrialisation and population growth have greatly increased the demand for potable water. Chennai City is considered as one of the important water stressed cities in the world. Regular water supply at adequate pressure is uncommon in the urban areas of developing countries. Intermittent supplies of piped water are the norm in most of the cities of Asia (UNEP, 1982). Full pressure water supply for 24 hours per day and 7 days in a week remains a pipedream in many of these cities (World Development Report, 2004) and Chennai is no exception. Table 4.13 shows water supply unreliability of some of the Asian cities.

It is evident that water supply unreliability is acute in Chennai City. Water unreliability affects productive activities,

Table 4.13. Water Supply Unreliability of Asian Cities

Cities	*Hours*	*Percentage of water unaccounted for*
Karachi	3	30
Delhi	4	25
Chennai	4	20
Colombo	22	28
Hong Kong	24	28
Bongkok	24	29
Beijing	24	9
Kolkatta	10	45

Source: Human Settlement Programmes, 2003.

business and Foreign Direct Investment (FDI) and also influences the mortality and morbidity of large sections of population.

Domestic Per Capita Supply

The prescribed water supply standard for Indian cities is the provision of at least 135 lpcd for domestic use, and this is based on various factors like climatic conditions, habits of people, water availability, water sources for non-drinking purposes, pressure in the pipeline, etc. The domestic water supply in the major cities of India is presented in Table 4.14.

The per capita water supply availability is the lowest in Chennai City. Due to the low per capita water supply, people try to find alternative sources like ground water and packaged drinking water. Most of the households possess dug well or bore-well to meet the non-drinking purposes, and purchase water

Table 4.14. Domestic Water Supply in Indian Cities, 2003

Cities	*Consumption (lpcd)*
Agra	115
Allahabad	125
Bangalore	90
Bhopal	100
Mumbai	150
Kolkatta	190
Delhi	160
Gaya	205
Gwalior	90
Hyderabad	205
Kanpur	205
Lucknow	90
Chennai	59
Mathura	120
Patna	115
Pune	275
Varanasi	115

Source: CMWSSB –Master Plan.

from private vendors for drinking purposes. The poor people solely depend on the Metro Water supply to meet their basic needs. Almost all the slum localities possess common dug-wells to tap water for their supplementary needs. Many sources of water are at some distance away from the area of demand as far as urban areas are concerned. Due to lack of storage and distribution facilities, poor people spend a sizeable proportion of the day for fetching water.

Environmental Problem

Some environmental problems also surface due to acute shortage of water supply. The problems are such as conflicts on surface and ground water resources, escalating water demand and lack of conservation measures, slow rate of water resources augmentation, deterioration of water quality and reduction in the yield of heavily exploited aquifers, inadequacies in the treatment of water and sewerage in developing urban communities, inefficient methods of waste water treatment and solid waste disposal, and continuing rapid growth of population (Walced .K, 2002). The failure of monsoon and the subsequent poor water supply lead to groundwater exploitation. The impact of poor monsoon on groundwater recharge coupled with high extraction in such stress periods have led to seawater intrusion in Minjur belt in North Chennai area (Shobhnath, 2003). The result was emergence of diseases and epidemic in urban areas either from untreated drinking water or from polluted water, which enters the water supply network through back siphonage or seepage (Malcolm, 1999). This also affects the health, especially of the poor people, as they do not have any safe water supply provision in their own premises.

Water pollution affects both surface water and ground water. Man has steadily been dumping wastes into the waterways, polluting the lakes, streams, rivers and seas. The ground water also gets contaminated due to rapid industrial growth. Further, inadequate sewerage and lack of wastewater treatment facilities cause river water and other sources of drinking water supply excessively polluted. Increasing water pollution further increases incidence of water borne diseases that require wastewater treatment process (The World Bank, 1994). More than half of the total surface water is polluted because of untreated urban sewerage (Indian Development Report, 2002).

Sanitation Problem

Water supply and sanitation are the prime constituents of social overhead capital. Together with safe water, adequate sanitation is now viewed as a fundamental human right. Adequate sanitation is necessary in controlling diseases and improving the quality of life in societies. At the basic level, adequate sanitation should ensure safe disposal of human waste and provide adequate water for personal cleaning to prevent food contamination and health hazards.

Water supply and sanitation are on the top of the development agenda in most of the developing countries. Due to unsafe water, more than three million people die every year from water-related diseases, mostly in developing countries (IMF and WB, 2003). The WHO has assessed that due to unsafe drinking water and lack of sanitation, there are more than 5 million deaths per year in developing countries. In 1994, approximately 1.4 billion people had no access to drinking water the world over. Lack of hygiene and the consumption of unsafe water allegedly cause 80 per cent of diseases in developing countries. Insufficient and inadequate supplies of water are the greatest single barrier for a healthy population and a productive community. Water-borne and water–related diseases are responsible for high infant mortality, low life expectancy and a poor quality of life. So, the provision of adequate protected water supply is of prime importance for public health.

Unsafe water brings high infant and child mortality, results in poor health, loss of productivity and shortened life.

Further, 1.085 to 2.87 million deaths worldwide in 2004 was due to diarrhoeal diseases that can be attributed to unhealthy water and sanitation. Moreover, 90 percentage of these occur in children under five (Down to Earth, May-15, 2004). In Asia, 1 in 3 people lacks access to safe drinking water, 5,00,000 infants die each year from diarrhoeal diseases related to lack of adequate water supply (Felix Clodds, 2000). A recent UN report (1997) notes that at any given time, 50 per cent of the population in developing countries is suffering from water related diseases, caused either by infection or indirectly by disease-carrying organisms. Thus, the quality of water supply and sanitation influence economic development. Chennai City faces acute shortage of water. Hence, adequate provision of healthy sanitation has become inevitable.

Problems in Water Pricing

Metering of domestic water consumption in any Indian City is rarely taken seriously. Meters that are installed generally do not work for very long time due to high particulate matter in the mains, which blocks the clock mechanism or due to damage caused by surges, when pressure returns to the intermittent water supply or due to vandalism. Even where they work the tariff charges are very less and are either not collected or are not worth collecting. Disconnecting domestic consumers for failure to pay is extremely unusual, although this sanction is occasionally used with commercial customers. The low priority accorded to collection has a significant impact on the revenue base of water entities. There are evidences that most domestic consumers are willing and able to pay a much higher rate than the official price for water (J. Roy, 2004).

Inadequate Finance

The availability of finance for the necessary investment in infrastructure, technology, and strong governance will improve the quality and quantity of urban water supply. As a result of rapid growth of urbanisation, the urban areas face the problems of river pollution, depletion, ground water contamination, and generation of wastewater. So it requires huge amount of investment and treatment methods. For the developing countries, one of the major hindrances to secure water supply is lack of adequate finance for maintenance, repair and improvement. International agencies are reluctant to allocate sufficient funds for water distribution projects in developing countries, because of acute financial constraints and the issue of water pricing. It is now known that such funding decisions are decided on political reasons rather than on economic considerations. For a healthy water supply system, therefore, water-pricing policy should cover at least the cost of operation and maintenance (Okun.D.1998).

In a democratic country like India, policies of water pricing for domestic consumption are decided largely on political grounds. In many cities, the revenue from water sales does not even cover the running and distribution costs of the network (Water Lorch, 1982). Poor billing arrangements and water prices, well below the costs of provision, are the causes for decline in the revenue of water utilities. The WHO/UNICEF Assessment 2000 has identified inadequate cost recovery and inadequate operation and maintenance as the two of the principal constraints on the development of proper water supply mechanism. Mounting arrears, low billing and lack of collection stipulate poor financial conditions to the utilities. In order to fill up the gap between demand and supply of finance, the water agencies are constantly seeking subsidies. Moreover, there is also public apathy about the public utilities. The causes of distrust of public utility services are mismanagement, widespread corruption, and unreliability services. These have led to further rise in the prices of water and consequently, the water agencies are not able to mobilise financial resources from the consumers.

Impact on the Poor

The poor are more prone to the adverse impacts of unsafe drinking water and inadequate sanitation; nevertheless they pay more for water. Asian Development Bank (1993) field surveys consistently show that the poor spend disproportionately more of their income on potable water than more privileged sections of the community for whom piped water supplies are assured (The Global, 2000 Report).

Emergence of Private Water Market

In order to the fill the gap between the demand and supply of water, private market is allowed to function in Chennai City. The private suppliers draw underground water from the neighbouring areas of Chennai City. Private water companies supply drinking water in different sizes of containers from sachets to water barrels, at a price ranging from Rs.1.00 to Rs.35 respectively.

Coca Cola, for instance, sources a bulk of its Kinley brand of bottled water from MVR Mineral Water, a contract bottler with a factory in Athur village, 40 kilometres from Chennai. The bottler reports that 132,000 litres of groundwater are extracted, each day. Of this, only 100,000 litres are used for bottling purposes. The remainder is discharged as effluents or lost due to other processes. Meanwhile, companies like Coca Cola, Pepsi, Parle, Team and Nestle seem to have unhindered access to free water from the ground. A network of tanks in the neighbouring area that stored rainwater and recharged the groundwater aquifers serve as the primary source of water, by ensuring the city's water security. Most of the watersheds and water bodies have been turned into cesspools of garbage, sewage and plastic wastes. According to the Central Ground Water Authority, more than 80 percent of the city's groundwater resources are already being tapped. Water scarcity in the metropolis turns to water crisis every summer. It is estimated that the households of Chennai City pay nearly $.10 million (Rs. 500 million) to private water companies for 3.7 billion litres of potable water each month, to augment the inadequate supply delivered by the state run Metro Water Board. Moreover it was reported that there are 400 illegal water-packaging units operating in the city and its surroundings, according to the South Indian Packaged Drinking Water Manufacturers Association (SIPDWA) (The Hindu, May 27, 2000). The details of prices charged by private water market in Chennai City are shown in Table 4.15.

Table 4.15. Chennai Water Market

Quantity of Water Supplied Through Water Market	*Price (Rs.)*
5 million quarter-litre packets at 2 cents	Rs. 1 each packet
75,000 1-litre bottles at 20 to 25 cents	Rs. 10 to Rs.12 each
100,000 12-liter cans at 38-63 cents	Rs. 18 to Rs. 30 each
25,000 20-25 litre bubble-top containers from 80 cents	Rs. 38 upwards
10, 000 water tankers carrying 12,000 litres	Rs. 700 to Rs. 900 per tanker.

Source: South India Packaged Drinking Water Manufacturers Association and The Hindu, Demand for Canned drinking water increasing: May 27th, 2000.

Note: Cents means I cent consists of 100 packets.

By sinking deep bore wells and using powerful pumps to draw water, water-packaging companies have managed to privatise the aquifers, which are common property resources. Communities self-sufficient in water until recently are on the edge of water insecurity.

Measures Adopted in Managing Water Crisis

Several measures have been taken by CMWSSB to meet the water crisis in Chennai City.

Water from Cauvery

In 1966, field investigation was carried out to draw water flowing through an open lined gravity canal from Cauvery River. The proposed length of the canal connecting Cauvery to Chennai City was 451kms. The project was finalised at an estimated cost of Rs.248.50 crores to augment the supply by as much as 1000 mld (220 mgd). This project could not be implemented because of opposition from various farmers' organisations citing its adverse impact on the agricultural system and environment.

As an alternative to the Cauvery River project, the Veeranam Lake was chosen as the source for pumping water and conveying it over long distance, by gravity. In 1967, Rs.23 crores was sanctioned for the project of 180 mld pumping schemes from Veeranam Lake to the city covering a distance of 220 kms. The total capacity of Veeranam Lake is 1485 mcft. The project was given up due to contractual litigation, and pipeline leakage when under pressure. Since then, the water flow in Cauvery River has dwindled owing to holding back of all the flows from Karnataka until their reservoirs overflowed. This has led to interstate water dispute. Moreover, Veeranam Lake itself is getting dry during the summer months (A. Mohanakrishnan, 1990).

Under the Chennai Water Supply Augmentation Project–1 (CWSAP-1) (New Veeranam Project) by the CMWSSB, the Cauvery basin is used as the source for supplying 180 million litres of water per day to Chennai City. This project was announced in 2001-2002, at a total cost of Rs.720 crores inclusive of treatment, conveyance and distribution. Raw water is pumped into the Vadakuthu treatment plant through the pipeline for treatment and pumps 180 mld of treated water to the Break Pressure Tank at Kadampuliyur Ridge point. It is then conveyed to the water distribution system at Porur in Chennai and is distributed to the public through the network system (Policy Note, 2004).

Chennai Water Supply Augmentation Project – II (CWSAP-II)

The aim of the CWSAP – II is to tap water from proximate areas of Chennai City. The project consists of the construction of 4 new reservoirs at Thirukkandalam, Zamin Korattur, Tiruneermalai, Pallikaranai and Palar. It envisages the deepening and desilting of Madhavaram, Retteri, Korattur and Ambattur lakes. The total cost of the project is estimated at Rs.493 crores.

Water from River Basin

Studies were conducted in 1987 under a UNDP aided project for exploring the availability of water from Cooum and Adyar River. However, hydrogeological survey has indicated that there is a very low potential for availing from these basins.

Palar is located 60 km south of Chennai City. This can become a major source of surface water for Chennai City, where water tapping is done from Palar anicut. The Palar river drainage basin area is quite large and the aquifer is deeper, where ground water potential is considerable. As a result of extensive hydrogeological studies carried out in 1987-89, five aquifers were identified at Athipattu, Pullarambakkam, Pilapur, Manapakkam and Voyalur. Out of these, the first three are farther away from the city and hence more suitable for nearer local bodies. The remaining two aquifers are considered suitable for further exploitation and the net surplus potential over and above the present drawal has been assessed as 40 mld. This is a potential source to be considered for the use by settlements in the southern limits of CMA like Tambaram, Pallavapuram, etc.

Desalination Plants

Desalination is the process of removing salt content from water. Five major processes have been used for desalination 1) Ion exchange 2) Distillation 3) Freezing 4) Reverse Osmosis and 5) Electro-Dialysis. Ion exchange has normally been used as a polishing process for industrial feed water processes. Among the four other methods, distillation, reverse osmosis and electro dialysis methods have proved commercially viable for water treatment. However, there is no one best method for desalination. There are two desalination plants, one at Melakadumallur using the process of Reverse Osmosis and the other one at Nochivayal on the process of Electro Dialysis. The cost of reverse osmosis plant was approximately Rs.34 to 40 crores and the cost of electro dialysis plant was Rs.28 to 45 crores. In Chennai, the experience with desalination plants has been successful. CMWSSB has established one plant each at Nochikuppam, Velachery and Ayodhyakuppam and two units at Kasimedu. In these plants reverse osmosis technology has been adopted for desalination of water. The plants produce potable water from brackish groundwater to supply water to a population of 40000. Table 4.16 gives the details of Chennai desalination plants.

Table 4.16. Details of Chennai Desalination Plant

Items	*Particulars*
Number of desalination plants	5
Aggregate capital cost	Rs.1.73 crores
Plants now operated	4
Quantum of water produced	5 lakh lpcd
Cost of water	Rs.36 per 1000liter

Source: CMWSSB – 2004, and the Hindu, Tackling Water Scarcity: "Government Floats Company" 7th July – 2004.

Krishna Water Supply Project

In 1976 the Chief Ministers of basin States, viz., Maharastra, Andhra Pradesh and Karnataka mutually signed an agreement to spare each 5 tmc.ft., of Krishna water to meet the water requirements of Chennai City. The agreement was signed between Andhra Pradesh and Tamil Nadu in 1983. Original estimate of the Krishna water supply project was Rs.760 crores, the project cost enhanced to Rs.2190 crores in the year 1983. Of the total cost of the project Rs.1018 crores had been executed in Andhra Pradesh (AP), and Tamil Nadu had so far paid Rs.512 crores to Andhra Pradesh. Besides this, Tamil Nadu had carried out capital works at a cost of Rs.180 crores. After transmission and evaporation loss, 12tmc ft of water would be realized at the Tamil Nadu border. Of this, three-tmc ft came initially. Further, in addition to three reservoirs catering to the city, the Chembarampakkam tank is planned to store Krishna water (Pasumalaidevan, 2001).

To provide additional capacity to store water, the Full Tank Levels of the water reservoirs of Chennai City have been

Table 4.17. Storage Capacity of Water Reservoirs before and After Improvements

Tank	*Before Improvement (mcft)*	*Capacity After Improvement (mcft)*	*Additional Storage*
Poondi Reservoirs	2750	3460	710
Chembarambakkam	3120	3645	525
Redhills	2850	3300	450
Total	8720	10405	1685

Source: CMWSSB, 1994-95.

Table 4.18. Water Received from Telugu Ganga Project for Chennai City

Year	*Quantity received (tmc) Per Year*
1996	0.076
1997	2.292
1998	2.812
1999	1.830
2000	3.591
2001	3.437
2002	3.342
2003	-
2004	0.104

Source: CMWSSB, Annual Reports-1996-2004.

increased by 60 cmts and their capacities before and after improvements are shown in Table 4.17.

The flow of Krishna water supply in Tamil Nadu is intended solely for the purpose of augmenting water supply to Chennai City. The component works up to the terminal reservoir of Red Hills Lake and Chembarambakkam tank and this is being implemented by the PWD, while the works connected with treatment, transmission, and distribution are being implemented by CMWSSB.

The water received at Tamil Nadu zero point from Kandaleru Reservoir (AP) under the Telugu Ganga Project from 1996 is (Krishna Water Supply) presented in Table 4.18.

Even after investing a huge sum of money for the Krishna Water Project, the total quantity of water realised for drinking water purpose of Chennai city is insignificant at 0.104 tmc as against the agreement of 12 tmc of water in the year 2004.

Waste Water Treatment and Reuse

Metro water authority for the benefit of industries initiated the programme of reuse of treated water in order to meet their demands. Industries have set up their own tertiary treatment plants to treat the secondary treated sewage supplied by CMWSSB from the Kodungaiyur sewage treatment plant. The total capacity of the plants put up by the industries is 30 mld. At present CMWSSB is supplying about 23 mld of treated sewage (with BOD 50 mgli and SS.70 mgli) to industries. Out of 230 mld 11.5 mld is supplied to Chennai Petroleum Corporation Ltd, and 11.5mld supplied to Madras Fertilizers ltd. The following Table.4.19 gives an account of the extent of water treated.

CMWSSB initiated to supply raw sewage of about 7mld for the power plants situated in North Chennai. The Petro-

Chemicals and Ennore port are expected to come into operation within the next 4 years and their demand for water is expected to be 50 mld per day.

Table 4.19. Water Treatment

Water treatment Stations / Plants	*Quantity of Water treated (mld)*
Kilpauk water works	270
Redhills	300
Veeranam	180
Chembarambakkam	530

Source: CMWSSB - 2004.

Decentralisation of CMWSSB

The managerial decentralisation of CMWSSB is mostly structured around procedures rather than services. The Board relies on tax property collection and government support for the implementation of its schemes. Since 1990s, the Board has had the courage to promote more decentralised resources. It promotes an internal delegation of power through better service interface. The outsourcing has been mostly decided on human resource policy basis. The board has a clear-cut budget allocation to various heads of expenditures and identifying tasks, duties and rights of the organisation. Similarly, it integrates the functions of supply, planning, management, and control of rights over assets.

It shall be noted here that, in the distribution aspects, outsourcing concerns the operation side, and not the supply activity (including commercial and so on). Thus, proper integration between control rights and financial rights remains, for a better incentive in decision-making. But, the allocation of financial resource to various expenditure items is still mediated through the political inter-sectoral arbitration of the corporation.

Estimation of Future Water Requirement

The CMWSSB has estimated the water requirements of Chennai City on the basis of a supply of 150 lpcd, inclusive of unaccounted water. The details are given in the Table 4.20.

Table 4.20. Future Water Requirements of Chennai City 2001-2021

Year	*Population*	*Water Requirements*		*Total*
		City (mld)	*Industries (mld)*	
2001	4.21	631	200	831
2011	6.00	900	250	1150
2021	7.00	1050	300	1350

Source: CMWSSB, Annual Report-2001-02.

It is expected that the water requirements will be increased to 900 mld and 250 mld in 2011 for the city and industries respectively. The requirements of water for Chennai City for the year 2001 were estimated by the CMWSSB at 631 mld for city domestic purposes (including consumption of households, institutional and commercial purposes) and 200 mld for industrial purposes. Due to heavy pressure of population, the respective water demand of the city and industries will be increased to 1050 and 300 mld respectively in the year 2021. A contingency Plan has been formulated at an approximate cost of Rs.700 crores to tackle the unprecedented drinking water scarcity with a view to maintaining a minimum lifeline supply to the six million population of Chennai City and adjacent urban areas. This plan, envisages extraction of additional ground water from different locations, provision of minimum additional infrastructure to convey the water and systems to equitably distribute the available water to all the residents through water tankers.

Extraction of Ground Water from Neyveli

Also, the contingency plan envisaged a proposal to extract ground water from Neyveli aquifer along the CWSAP-I alignment by constructing deep bore wells for conveying 60 million litres of water every day for about 5 months from March to July every year, to supplement the additional requirement of water in Chennai City during summer. Accordingly, works were taken up to construct 45 deep bore wells between Paravanar River near Sethiathope and Gadilam River near Panruti, extending to a distance of about 30 km.

Tariffs and Cost Management

The CMWSSB has managed in 2000 to cover its operational expenses, with a net surplus of Rs. 25.5 crores. However, the capital expenditure still comes from or routed through the State Government, raising the issue of the long-term sustainability of the water supply and sanitation in Chennai. More precisely, in the absence of State Government support, from 1997, it goes on depending on the World Bank loans. The Board is presently considering an indexation of tariffs. As of now, the Board is close to the functioning of an administrative body as far as revenues are concerned. Indeed, a noticeable part of the 'operational' revenue (27 per cent) comes from the collected tax, which is rather an administrative prerogative than something correlated to water supply or sanitation. As far as financial results are concerned, the increase in the tax value explains 64 per cent of the total increase in 'operational revenue'. In 1999-2000, the State Government has granted Rs.165 crores, while total loans from the Government amounted to Rs.656 crores, and loans from financial institutions only by Rs.129 crores (CMWSSB, Annual Report 2000-01). Overall efficiency in collecting taxes is varying year wise between 80 and 90 per cent, while bills are collected with a level of 80 per cent only. As far as water supply is concerned, the Board gets more than half of its water operational income from 40 big consumers only.

Water supply system has expanded in an unplanned manner. Chennai City faces acute water scarcity, because demand always exceeds the supply. This leads to drastic depletion of ground water. Planners project future demand, without considering whether the available supplies could be sustainable over a long period. Water institutions, laws, government agencies, planning and engineering concentrate mainly on supply side management. The conventional approach of supply side management may work only when the water is abundant and it is not suitable to a period of water scarcity and rising water demand. The following factors are responsible for the weakening of the water supply sector.

5

Socio-cultural Profiles of Sample Households and Housing Characteristics

The present chapter is concerned with the socio-cultural profiles of the sample households in Chennai City. The topics covered are the demographic particulars and general characteristics of the sample households and the aggregate of their membership, economic condition of households, housing conditions with respect to various basic and common facilities, education and cultural interests. To be more specific, the entire chapter has been divided into several sections, each dealing with one aspect of the particular study.

The household is the sampling unit of the survey, which stands for the family. It is the basic social unit. Some social and economic conditions are common to a household; however, some others are different for different households. Among them, the head occupies the most important position and, therefore, it is useful to acquaint with the heads of the sample households who are the respondents of the study.

Characteristics of the Heads of the Households

The heads of the households have been distributed with some characteristics.

SEX

Table 5.1. Sex of Heads of the Households

Sex	*Location*		*Total*
	North Chennai	*South Chennai*	
Male	201 (89.3)	205 (91.1)	406 (90.2)
Female	24 (10.7)	20 (8.9)	44 (9.8)
Total	225 (100.0)	225 (100.0)	450 (100.0)

Source: Survey Data, 2005.
Note: Figure in the parenthesis indicates row percentage.

Both male and female heads are represented in the study area. Table 5.1 presents the sex-wise distribution of heads of the households. To take up sex first, 44 or 9.8 per cent of the 450 heads in the sample households are females. The proportion of female head households in North Chennai is 10.7 per cent, when compared to 8.9 per cent in South Chennai. This shows that almost one out of every ten heads is a female.

Figure 5.1.Sex of Heads of the Households

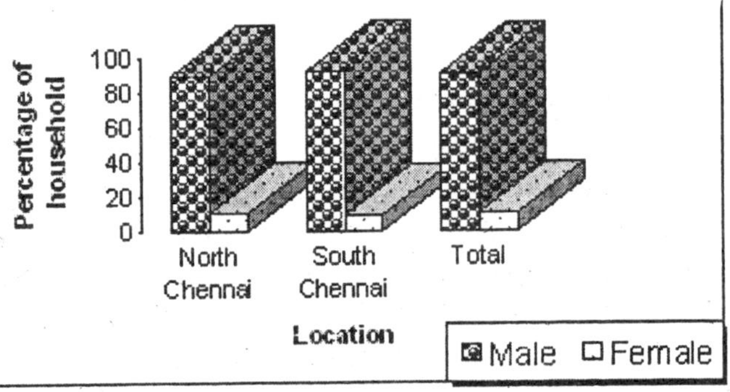

Table 5. 2. Age Distribution of Heads of Households

Age (in years)	*Sample households*	*Percentage*	*Minimum*	*Maximum*	*Mean*	*Std. Deviation*
Below 24	7	1.6	22.00	24.00	23.4286	.78680
25 - 39	101	22.4	25.00	39.00	33.2673	3.64662
40 - 59	236	52.4	40.00	59.00	47.8432	5.72386
60 and above	106	23.6	60.00	82.00	66.7075	6.10381
Total	450	100.0	-	-	-	-

Source: Survey Data, 2005.

AGE

To differentiate among the heads by age, the percentage share of age groups in the total number of heads is shown in Table 5.2.

The age group claiming the largest share is those heads aged between 40 to 59 years of age. The age group 40 to 59 years accounts for a little over half (52.4 per cent) the heads with a mean age of 47.84 years and standard deviation of 5.7238. Correspondingly the proportion of aged heads forms 23.6 per cent with a mean age of 66.7 years. Relatively speaking the older heads administer the households more in numbers as compared to the younger ones. The younger generation heads aged between 25 and 39 years represent 22.4 per cent of the total households with a mean age of 33.27.

The age distribution of heads of households according to the location is displayed in Table 5.3.

The important inference derived is that relatively younger people head the households in North Chennai when compared to the households in South Chennai.

Table 5.3. Age Distribution of Heads of Households

Age (in years)	*North Chennai*		*South Chennai*	
	Sample households	*Percentage*	*Sample households*	*Percentage*
Below 24	4	1.8	3	1.3
25 - 39	57	25.3	44	19.6
40 - 59	119	52.9	117	52.0
60 and above	45	20.0	61	27.1
Total	225	100.0	225	100.0

Source: Survey Data, 2005.

HOUSEHOLD CHARACTERISTICS

The survey of the sample households has shown that there are 1945 members in 450 households with an average household size of 4.32 persons. This is considerably lesser than the all India average of 4.7 members per household. Since household is the basic unit of any economy or civil society, some of the characteristics are common to all the households and yet each household differs from the other in some respects. Such characteristics are consolidated under a common head as household characteristics.

COMMUNITY

The sample households fall into different castes and communities. For the sake of simplicity, large number of caste classifications is considered under different communities. This could be seen in Table 5.4.

It is observed that a majority of the households is Hindus followed by Christians and Muslims. All these religious groups are classified under Scheduled Castes, Scheduled Tribes, Most Backward Castes, Backward Castes and Others. The Backward Caste households are the largest of the five groups, both in North Chennai and South Chennai. On the whole, they represent nearly half (49.8 per cent) of the total households. The percentage variations of communities between North Chennai and South Chennai are very meagre. The second

Table 5.4. Community of the Households

Community	*Location*		
	North Chennai	*South Chennai*	*Total*
Scheduled Caste	76 (33.8)	50(22.2)	126 (28.0)
Scheduled Tribe	6 (2.7)	4(1.8)	10 (2.2)
Most Backward Caste	25 (11.1)	34(15.1)	59(13.1)
Backward Caste	113 (50.2)	111 (49.3)	224(49.8)
Others	5 (2.2)	26 11.6)	31(6.9)
Total	225(100.0)	225(100.0)	450(100.0)

Source: Survey Data, 2005.

Note: Figure in the parenthesis indicates column percentage.

largest community living in Chennai City is the Scheduled Castes. They account for 28 per cent of the sample households, followed by the Most Backward Castes representing 13.1 per cent. A good degree of similarity exists between North Chennai and South Chennai with regard to the ranking of the above-cited three communities; however, other communities register their presence five times more in South Chennai as compared to North Chennai.

LITERACY STATUS

Literacy status is generally perceived as the ability of a person to read and write a given language. The aggregate responses of literacy status of the heads in North Chennai and South Chennai are shown in the Table 5.5.

Table 5.5. Literacy Status of Heads of Households

Literacy status	*Location*		*Total*
	North Chennai	*South Chennai*	
Literate	167 (74.2)	188 (83.6)	355 (78.9)
Illiterate	58 (25.8)	37 (16.4)	95 (21.1)
Total	225 (100.0)	225 (100.0)	450 (100.0)

Source: Survey Data, 2005.
Note : Figure in the parenthesis indicates column percentage.

The overall proportion of literate heads accounts for 78.9 per cent. The proportion of literate heads in North Chennai is relatively smaller as compared to South Chennai. It indicates that households in North Chennai are poorer and constitute middle class households.

Further information was sought concerning the literacy level of individual head of the household. All the literate heads reported that they had received some formal education. Based

Table 5.6. Educational Standard

Educational standard	*Location* North Chennai	South Chennai	*Total*
Primary (I to V Std)	12	7	19
	(7.2)	(3.7)	(5.4)
Middle (VI to VIII Std)	40	39	79
	(24.0)	(20.7)	(22.3)
High school (IX to X Std)	66	73	139
	(39.5)	(38.8)	(39.2)
Higher secondary / PUC (XI, XII & PUC)	20	20	40
	(12.0)	(10.6)	(11.3)
Graduate	17	30	47
	(10.2)	(16.0)	(13.2)
Post-graduate	9	13	22
	(5.4)	(6.9)	(6.2)
Professional	1	5	6
	(.6)	(2.7)	(1.7)
Others	2	1	3
	(1.2)	(.5)	(.8)
Total	167	188	355
	(100.0)	(100.0)	(100.0)

Source: Survey Data, 2005.
Note : Figure in the parenthesis indicates column percentage.

on their responses, the educational standards of the heads are furnished in Table 5.6.

It is evident that nearly two-fifths (39.2 per cent) of the literate heads are educated up to High School level. Taking all educated heads as a whole in the sample, it can be understood that the graduates and postgraduates are in the percentages of 13.2 and 6.2 respectively. The details of literacy status could not present the gulf in the educational standards of the respondents. But it is inferred that heads in South Chennai record a much better showing in higher education, when compared to the heads in North Chennai.

Household Composition

Except uni-member households, the household represents a group of persons who share food from one kitchen. The institution of family or household is defined as a group of persons having kinship or blood ties with one another, having one oven for cooking food.

Type of Households

Based on the content of relationship, the families are classified into uni-member, nuclear family and joint-family. Uni-member household is a single person living in a house whereas, nuclear and joint-families are classified on the basis of structure and internal relationships between members of family. Nuclear families represent husband, wife and unmarried children, while joint family includes other types of blood relatives as well. According to this scheme, the sample households are classified under three heads. The percentage distribution of different types of families is shown in Table 5.7.

It is very conspicuous that uni-member families are very rare and they represent 0.9 per cent of the total households

Table 5. 7. Type of Families

Type of family	*Location*		*Total*
	North Chennai	*South Chennai*	
Uni-member	2 (0.9)	2 (0.9)	4 (0.9)
Nuclear	172 (76.4)	185 (82.2)	357 (79.3)
Joint	51 (22.7)	38 (16.9)	89 (19.8)
Total	225 (100.0)	225 (100.0)	450 (100.0)

Source: Survey Data, 2005.
Note : Figure in the parenthesis indicates column percentage.

Table5. 8. Average Size of Households

Size of the household	*Sample households*	*Percentage*	*Minimum*	*Maximum*	*Mean*	*Std. Deviation*
Below 2	50	11.1	1.00	2.00	1.9200	.27405
2 - 4	214	47.6	3.00	4.00	3.6869	.46484
4 - 6	154	34.2	5.00	6.00	5.3506	.47873
6 - 8	28	6.2	7.00	8.00	7.1071	.31497
8 and above	4	0.9	9.00	10.00	9.2500	.50000
Total	450	100.0	1.00	10.00	4.3222	1.44712

Source: Survey Data, 2005.

Figure 5.3. Size of Households

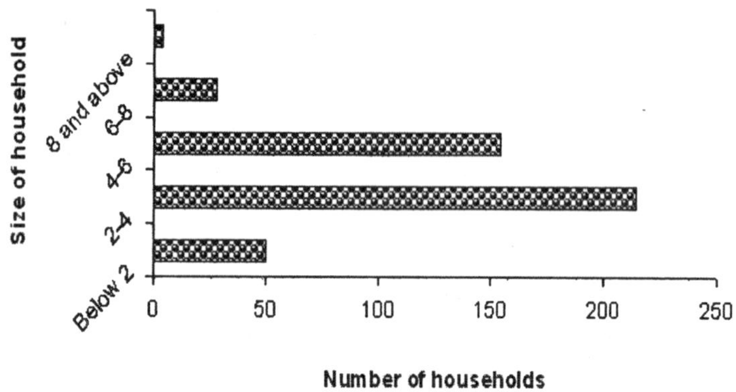

surveyed. The nuclear households form the largest group, accounting for 79.3 per cent. The preponderance of nuclear families is of much higher degree in South Chennai. The joint families accounted for the sample as a whole form 19.8 per cent.

SIZE OF HOUSEHOLD

The average size of a household is nothing but the number of members in a household. The overall variations in the number of members range from one to as many as nine members in a household. Table 5.8 presents the descriptive of the distribution of households in different groups of household size.

It is found that 214 of the 450 households (47.6 per cent) have two to four members in the family, with a mean family size of 3.7. The next biggest household size is four to six members presenting 34.2 per cent of the total sample households having a mean size of 5.4. These two groups of household's collectively form 81.8 per cent or four-fifths of the total sample households.

DEMOGRAPHIC COMPOSITION

The descriptive statistics of demographic composition of the sample households is shown in Table 5.9. It is found that

Table 5. 9. Descriptive Statistics of Demographic Variables

Demographic variables	*N*	*Minimum*	*Maximum*	*Mean*	*Std. Deviation*
Total number of male members	445	1.00	7.00	2.2135	1.06619
Total number of female members	448	1.00	6.00	2.1429	1.03069
Total number of children (below 5 years)	89	1.00	4.00	1.6404	0.67835
Total number of children (between 5 and 15 years)	119	1.00	4.00	1.6807	0.68807
Total number of members (above 15 years)	450	1.00	9.00	3.5533	1.45857
Number of income earners	450	1.00	5.00	1.4511	0.72102
Total number of married persons	445	1.00	8.00	2.4090	0.97902

Source: Survey Data; 2005.
Note : N- denotes Number of Households.

445 of the 450 households (98.8 per cent) have male members ranging from one to seven having a mean size of 2.21. There are 99.5 per cent (448 out of 450) households with female members ranging their size from one to six with the mean size of 2.14. On an average the number of male members is higher than that of the female members. The children are classified into non-school-going aged children and school-going-aged-children. The non-school going children are aged below 5 years and they are found among 19.7 per cent (89 out of 450) households, with an average size of 1.64. The school going children aged between 5 and 15 years are represented in 26.4 per cent (119 out of 450) households.

The average number of working age population in an average household, inclusive of the aged members, is 3.55. Of them, 1.45 members per households are in the category of income earners. While analyzing the marital status of the household members in the marriageable age, it is learnt that the married persons range from 1 to 8 persons in 445 households with the mean value of 2.4.

Economic Condition

The discussion of economic condition centers round the concept of household income, because it explains the ability of a household to demand goods and services to secure a given or higher standard of living. Household income is the total of all types of earnings received by the members of each household in economically gainful activities. A statement of the monthly income of sample households is given in Table 4.10.

Aggregated monthly incomes of all households give a sum of Rs.37.66 lakhs. The frequency distribution of monthly household income explains that 29.1 per cent of the households earn an income between Rs. 2500 and Rs.5000. Another 19.3 per cent households earn less than Rs.2500 per month. Thus nearly half (48.4 per cent) of the sample households earn less

Table 5. 10. Monthly Income of the Households

Income levels (in Rs.)	*S.HHs*	*%*	*Minimum (Rs.)*	*Maximum (Rs.)*	*Mean*	*Std. Deviation*	*Gross Income*
Below 2500	87	19.3	300.00	2500.00	1932.03	491.18	168087
2500 - 5000	131	29.1	2700.00	5000.00	3822.14	750.27	500701
5000 - 7500	63	14.0	5190.00	7500.00	6314.40	679.59	397807
7500 - 10000	47	10.4	7850.00	10000.00	9152.30	768.44	430158
10000 - 12500	12	2.7	10100.00	12500.00	11216.67	768.51	134600
12500 - 15000	33	7.3	12560.00	15000.00	14110.61	819.15	465650
15000 and above	77	17.1	15100.00	80000.00	21677.40	7944.97	1669160
Total	450	100.0	300.00	80000.00	8369.25	7668.79	3766163

Source: Survey Data, 2005.

Note : S.HHs- indicates Sample households and % shows percentage.

Figure 5.4. Monthly Income of the Households

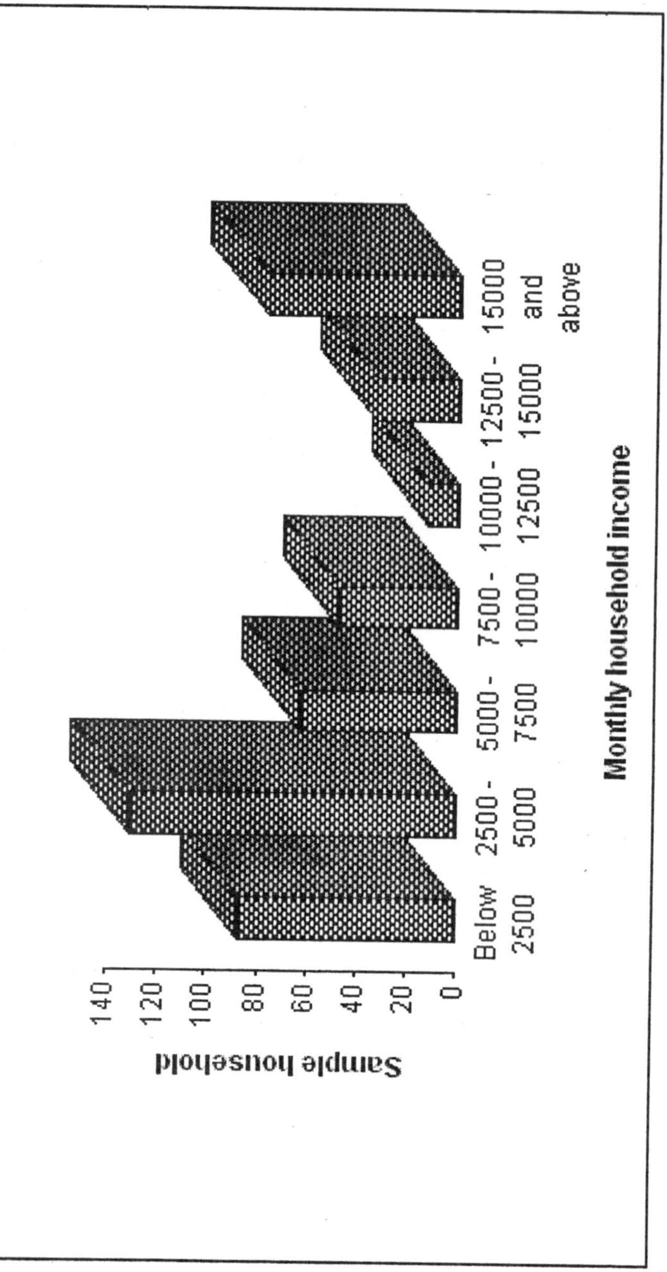

Table 5. 11. Distribution of Monthly Household Income

Monthly Household Income (Rs.)			*Location North Chennai*			
	No.of House-holds	*Minimum (Rs.)*	*Maximum (Rs.)*	*Mean*	*Std. Deviation*	*Gross Income (Rs.)*
Below 2500	68 (30.2)	300.00	2500.00	1881.13	499.99	127917
2500 - 5000	81 (36.0)	2700.00	5000.00	3796.10	763.12	307484
5000 - 7500	33 (14.7)	5250.00	7500.00	6264.15	724.09	206717
7500 - 10000	12 (05.3)	8000.00	10000.00	8944.42	633.02	107333
10000 - 12500	3 (01.3)	11000.00	12500.00	11833.33	763.76	35500
12500 - 15000	11 (04.9)	13000.00	15000.00	14090.91	943.88	155000
15000 and above	17 (07.6)	16000.00	29000.00	21323.53	3916.88	362500
Total	225 (100.0)	300.00	29000.00	5788.6711	508.67564	1302451

(Contd.)

(Contd. Table 5.11)

Monthly Household Income (Rs.)	*Location South Chennai*						*Grand Total of Households*
	No. of Households	*Minimum (Rs.)*	*Maximum (Rs.)*	*Mean*	*Std. Deviation*	*Gross Income (Rs.)*	
Below 2500	19 (8.4)	1370.00	2500.00	2114.21	420.92	40170	87 (19.3)
2500 - 5000	50 (22.2)	2750.00	5000.00	3864.34	734.63	193217	131 (29.1)
5000 - 7500	30 (13.3)	5190.00	7500.00	6369.67	634.67	191090	63 (14.0)
7500 - 10000	35 (15.6)	7850.00	10000.00	9223.57	805.47	322825	47 (10.4)
10000 - 12500	9 (4.0)	10100.00	12000.00	11011.11	690.01	99100	12 (2.7)
12500 - 15000	22 (9.8)	12560.00	15000.00	14120.45	773.26	310650	33 (7.3)
15000 and above	60 (26.7)	15100.00	80000.00	21777.67	8780.86	1306660	77 (17.1)
Total	225 (100.0)	1370.00	80000.00	109.4983	8611.58	2463712	450 (100.0)

Source: Survey Data, 2005.
Note : Figure in the parenthesis indicates column percentage

than Rs.5000 per month. It is found that 17.1 per cent of the sample households earn Rs.15000 and above per month with an average income of Rs.21, 677.4 and a standard deviation of 7944.97.

The distribution of monthly household income according to the location of households is shown in Table 5.11.

It can be seen that the gross income received by 225 households in North Chennai is almost half of the gross income received by households in South Chennai. While making a comparison of the mean income in each income group, it is clear that in all the income groups, except in the range of Rs.12500-Rs.15000, the mean income of South Chennai is higher than the mean income of North Chennai. The magnitude of income inequality may be better understood if the frequency distribution of household income of the two locations is compared. It is interesting to note that 30.2 per cent of the sample households in North Chennai earn a monthly household income below Rs.2500, when compared to the 8.4 per cent of the same in South Chennai. At the other extreme, 26.7 per cent of the sample households in South Chennai earn a monthly household income of Rs.15000 and above as compared to just 7.6 per cent of the same in North Chennai. The households in South Chennai appear to be much better off economically than the households in North Chennai.

Housing Characteristics

The housing characteristics of the sample households exhibit some variations. Deficiencies and accentuation of allied amenities and services have not been equal, but varies in degree. Aspects of data on housing characteristics are based on the spot appraisal of the prevailing conditions. Against this background the data about the internal facilities enjoyed by the sample households and the local conveniences available to them are described.

Nature of Residency

Information about the ownership of houses is obtained for all the sample households. The information is furnished in Table 5.12.

It is found that 69.8 per cent of the households owned the houses occupied by them. The proportion of houses owned and occupied by the households in North Chennai and South Chennai vary very less. There are 2.7 per cent households each in South Chennai and North Chennai who live in Government Quarters. The type of settlements of the sample households may be

Table 5. 12. Nature of Residency

Nature of residency of the household	*Location*		*Total*
	North Chennai	*South Chennai*	
Owned - Occupied	153(68.0)	161(71.6)	314(69.8)
Rented	66(29.3)	58(25.8)	124(27.6)
Government Quarters	6(2.7)	6(2.7)	12(2.7)
Total	225(100.0)	225(100.0)	450(100.0)

Source: Survey Data, 2005.
Note : Figure in the parenthesis indicates column percentage.

Table 5. 13. Types of Houses

Type of Settlements	*Location*		*Total*
	North Chennai	*South Chennai*	
Independent house	134 (59.6)	136 (60.4)	270 (60.0)
Independent storeyed house	72 (32.0)	67 (29.8)	139 (30.9)
Apartments	18 (8.0)	22 (9.8)	40 (8.9)
Others	1 (0.4)	-	1 (0.2)
Total	225 (100.0)	225 (100.0)	450 (100.0)

Source: Survey Data, 2005.
Note : Figure in the parenthesis indicates column percentage.

classified into independent houses, independent storeyed houses, apartments and others.

Table 5.13 classifies the categories of houses occupied.

With regard to the type of houses, the data indicate that three-fifths (60 per cent) of the households live in independent houses; some 30.9 per cent live in independent storeyed buildings and another 8.9 per cent by apartments. However, the types of houses do not really reveal their structure. It is therefore, interesting to assess the physical structure of the houses occupied by them.

Physical Structure of Houses

For an easy appraisal of the physical characteristics of these houses, they are differentiated on the basis of the roof structure. These differentiations relate to (1) the nature of structure (2) type of construction and (3) the kind of roof. The frequency distribution of the physical structure of the houses based on these categories is shown in Table 5.14.

Table 5. 14. Physical Structure of Houses

Nature of construction	*Location*		*Total*
	North Chennai	*South Chennai*	
Luxury bungalow	11(4.9)	19(8.4)	30(6.7)
Concrete roof	167(74.2)	181(80.4)	348(77.3)
Tiled roof	15(6.7)	15(6.7)	30(6.7)
Thatched roof	28(12.4)	10(4.4)	38(8.4)
Others	4(1.8)	-	4(0.9)
Total	225(100.0)	225(100.0)	450(100.0)

Source: Survey Data, 2005.
Note : Figure in the parenthesis indicates column percentage.

In both North Chennai and South Chennai, the common type of the house, which is of concrete roof, claims over three fourths (77.3 per cent) of the total houses. Houses of luxurious

Bungalow type constitute 6.7 per cent. Most of the houses of this type are from South Chennai. In both locations majority of the poor households live in slums, in tiled roof and thatched roof dwelling units. Such households are in the proportion of 6.7 and 8.4 percentages respectively. However, the typical house in Chennai City is represented by concrete roofs.

Area of Houses

Excepting the dwellings in slums or huts, the floor size of the houses reveals quite an extensive range. The size of the floor area of houses of the sample households is shown in Table 5.15. The descriptive data on the floor area of houses occupied by the sample households reveal that the average floor area of a dwelling unit is computed as 877.11 sq.ft. The floor areas of houses of the sample households range from 100 sq.ft., to a maximum of 2500 sq.ft.

The comparison of floor area of houses between North Chennai and South Chennai reveal that the variation has been very small; however, average floor area of houses in South Chennai is larger than those in North Chennai.

Table 5. 15. Floor Area of Houses

Total Floor Area of Residence	*Location*		*Total*
	North Chennai	*South Chennai*	
Number of houses	225	225	450
Minimum (sq.ft)	40.00	100.00	100.00
Maximum (sq.ft)	2500.00	2000.00	2500.00
Mean (sq.ft)	808.22	946.00	877.11
Std. Deviation	397.00	444.91	426.78
Gross Area (sq.ft)	181850	212850	394700

Source: Survey Data, 2005.

Number of Rooms

The question of living accommodation depends upon the number of living rooms in a house. Living rooms are treated as the areas of physical separation of one room from the other, whether it is a living room or bedroom, drawing or sitting room, except kitchen. In this connection it may be useful to mention that the slum dwellings with a single room are all that is available for all the living purposes. The availability of rooms in the houses of the sample households is presented in Table 5.16.

Table 5. 16. Rooms of the house

Number of Rooms in the house	*Location*		*Total*
	North Chennai	*South Chennai*	
One	58(25.8)	31(13.8)	89(19.8)
Two	66(29.3)	76(33.8)	142(31.6)
Three	75(33.3)	76(33.8)	151(33.6)
Four	20(8.9)	36(16.0)	56(12.4)
More than four	6(2.6)	6(2.7)	12(3.3)
Total	225(100.0)	225(100.0)	450(100.0)

Source: Survey Data, 2005.
Note : Figure in the parenthesis indicates column percentage

Interestingly, nearly one-fifths (19.8 per cent) of the households live in single room tenements. Nevertheless, households who live in houses with more than four rooms account for 2.7 per cent. The distribution of households according to the availability of rooms is more or less similar for both North Chennai and South Chennai. However, single room houses are a little more than one-fourth (25.8 per cent) in North Chennai as compared to 13.8 per cent of the same in South Chennai.

Building Complex

To get an idea of the prevailing congestion within dwellings, and for a rough idea of living floor space, it has been decided to ascertain the actual number of households including the sample household live in each building where the household is occupied. The available information on this point is furnished in Table 5.17.

Table 5. 17. Number of Households staying in the Building

Households in a having complex	*Location*		*Total*
	North Chennai	*South Chennai*	
One	198(88.0)	189(84.0)	387(86.0)
Two	20 (8.9)	23(10.2)	43 (9.6)
Three	6 (2.7)	8 (3.6)	14 (3.1)
More than three	-	2 (0.8)	2 (0.4)
Total	225(100.0)	225(100.0)	450(100.0)

Source: Survey Data, 2005.

Note: Figure in the parenthesis indicates column percentage.

It is observed that, even in slums, more than one household live in the same dwelling unit. There are some buildings in non-slum areas where it is quite common to have more than one household in one house building. Among the sample households it is found that 86 per cent live as a single unit household. Houses with two and three households are in the proportion of 9.6 and 3.1 percentages respectively. Accommodation of more than three households in the same building accounts for 0.4 per cent only.

Internal Facilities

Many of the housing units have some essential facilities such as a lavatory (flushing toilets), both showers, kitchen sink, bath rooms and wash basins. All these facilities require sufficient supply of water. These facilities are shared among the members in a household. The sample households with different types of housing facilities are shown in Table 5.18.

It is generally perceived that the availability of these internal facilities depends upon adequate supply of water from public or private sources. Moreover, the number of these

Table 5.18. Different Types of Housing Facilities

Nos	*Number of showers Location*			*Number of flushing toilets Location*		
	N C	*SC*	*Total*	*N C*	*SC*	*Total*
One	42 (95.5)	69 (95.8)	111 (95.7)	181 (86.2)	105 (48.4)	286 (67.0)
Two	2 (4.5)	3 (4.2)	5 (4.3)	25 (11.9)	104 (47.9)	129 (30.2)
Three				3 (1.4)	6 (2.8)	9 (2.1)
Four				1 (.5)	2 (.9)	3 (.7)
Total	44 (100.0)	72 (100.0)	116 (100.0)	210 (100.0)	217 (100.0)	427 (100.0)

(Contd.)

(Contd. Table 5.18)

Nos	Number of kitchen sink Location			Number of bath rooms Location		
	N C	SC	Total	N C	SC	Total
One	69 (93.2)	85 (57.8)	154 (69.7)	192 (87.7)	105 (47.9)	297 (67.8)
Two	5 (6.8)	58 (39.5)	63 (28.5)	22 (10.0)	103 (47.0)	125 (28.5)
Three		2 (1.4)	2 (.9)	4 (1.8)	9 (4.1)	13 (3.0)
Four		2 (1.4)	2 (.9)	1 (.5)	2 (.9)	3 (.7)
Total	74 (100.0)	147 (100.0)	221 (100.0)	219 (100.0)	219 (100.0)	438 (100.0)

Source: Survey Data, 2005.
Note: Figure in the parenthesis indicates column percentage.

internal facilities approximately determines the demand for water for domestic use. The availability of internal facilities increases with the extent of living accommodation. In the context of the number of showers, it is found that 116 of the 450 households (25.7 per cent) possess showers in their bathrooms. A majority of the sample households (95.7 per cent) are connected with single shower. The distribution of showers in the households across North Chennai and South Chennai exhibits almost the same level. No household possesses more than two showers in both North Chennai and South Chennai.

As for the availability of flushing toilets, 23 households, out of the 450 (5.1 per cent) do not posses a flushing toilet. Among which those dwellings that enjoy toilet facility, 67 per cent have a single toilet, 30.2 per cent have two toilets; but the households with three and four toilets account for 2.1 and 0.7 percentages respectively. The comparison in the distribution of toilets across the two locations reveals that North Chennai households lack this facility much more when compared to the households in South Chennai. Nearly half of the total households (49.1 per cent) possess the facility of kitchen sink. Only 74 households (16.4 per cent) in North Chennai have kitchen sink when compared to 65.3 per cent (147 out of 225) of the same in South Chennai. It means that nearly one out of every four households possesses kitchen sink in South Chennai. If we consider the number of kitchen sinks in the sample households, it can be understood that 93.2 per cent of the households possessing kitchen sinks have only one sink in North Chennai, when compared to 57.8 per cent of the same in South Chennai.

As for the number of bathrooms, 438 of the 450 households (97.3 per cent) have bathrooms. It is usual for the slum-dwellers to posses a make shift bathroom covered by thatched coconut leaves. If households with bathrooms are taken for computation then, 87.7 per cent in North Chennai and 47.9 per cent in South Chennai have single bathrooms If a household has more than one bathroom, it will be self-contained within the dwelling units. Thus, 47 per cent of such households in South Chennai have two bathrooms, when compared to 10 per cent of similar households in North Chennai.

It is found that 201 out of 450 households (44.67 per cent) have washbasins in their dwellings. Households in South Chennai enjoy such facilities much more than households in North Chennai. If this facility is taken for South Chennai separately, 55.8 per cent of such households have single washbasin and 42.8 per cent with two washbasins.

For an overall view, it may be noted that the largest group of dwellings with one room accommodation lack shower, kitchen sink and washbasins. Bathrooms and toilets are there in a larger proportion of the sample households.

Other Water Conveniences

Adequate water supply and its accessibility are commonly accepted as essential for human life. With regard to urban water supply, the quality of service may be limited to the source from which the needs of the households are met. Among the different sources, public water supply from CMWSSB is by far the most predominant. Public water supply to the households is provided either through piped water connection or through non-piped water supply. The piped water connection may be metered or un-metered. Some other households are supplied with in-house water connection with only a hand pump. Some households are supplied water through pipes directly to the sump.

Table 5.19 depicts the types of in-house water connection available to the households.

It can be noticed that nearly three-fifths (59.1 per cent) of the households in South Chennai have in-house connection as

Table 5.19. Types of In-house Water Connection

In-house connection	*North Chennai*	*South Chennai*	*Total*
Metered in-house connection	-	3 (1.9) [100.0]	3 (1.1) [100.0]
Un-metered in-house connection	14 (13.0) [30.4]	32 (20.5) [69.6]	46 (17.4) [100.0]
In-house connection with only hand pump	94 (87.0) [43.7]	121 (77.6) [56.3]	215 (81.4) [100.0]
Total	108 (100.0) [40.9]	156 (100.0) [59.1]	264 (100.0) [100.0]

Source: Survey Data, 2005.
Note : Figure in the parenthesis indicates column percentage.
Figure in the brackets indicates row percentage.

compared to 40.9 per cent households in North Chennai. It is found that totally metered water supply is available only to 1.1 per cent of the households with in-house water connection. More than four-fifths of the households with in-house water connection have only the hand-pump facility.

Un-metered in-house connection with sump facility is available for 17.4 per cent of the households. The particulars of public water supply in common distribution points are available in Table 5.20.

A majority of the households (68.6 per cent) depend on common water sources. Households fetch water from public hand pump, public street tank and mobile tanker sources are in the percentages of 68.6, 12 and 16 respectively. The important inference is that more households (75.2 per cent) in North Chennai depend on public hand pump as compared to 56.5 per cent of the same in South Chennai.

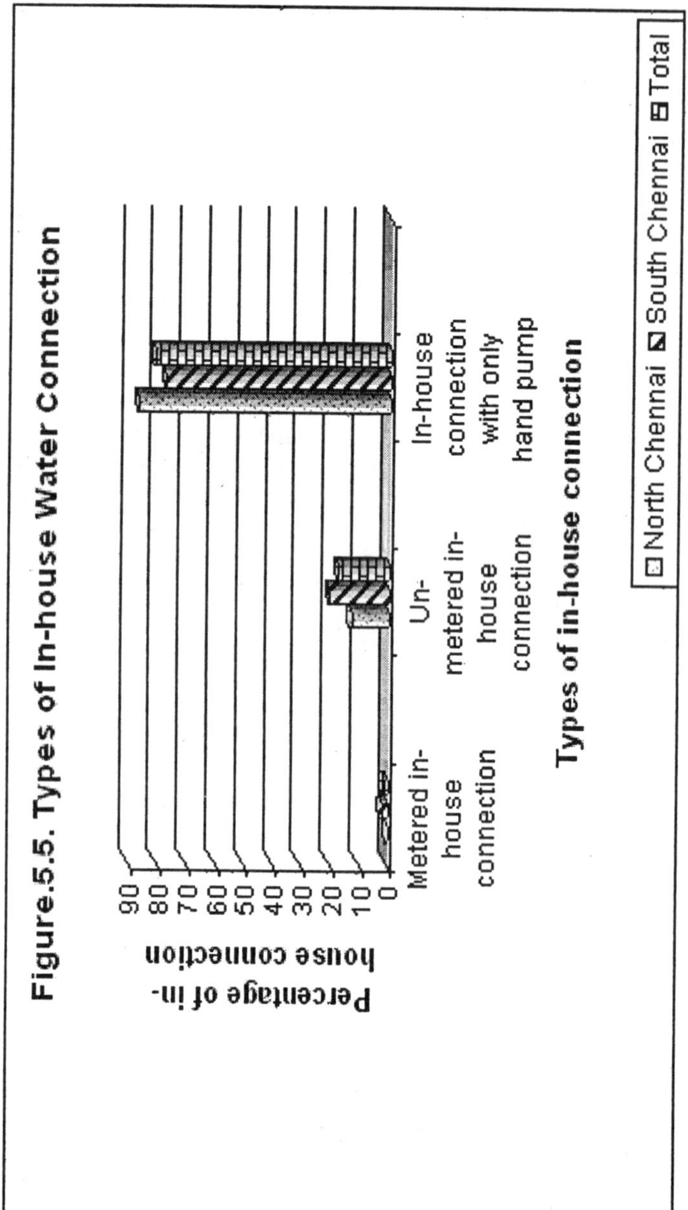

Figure.5.5. Types of In-house Water Connection

Table 5.20. Types of Public Water Supply in Common Distribution Points

Types of public water supply	*North Chennai*	*South Chennai*	*Total*
Public water tap	5 (4.4)	1 (1.6)	6 (3.4)
Public hand pump	85 (75.2)	35 (56.5)	120 (68.6)
Public street water tank	8 (7.1)	13 (21.0)	21 (12.0)
Mobile tanker lorry	15 (13.3)	13 (21.0)	28 (16.0)
Total	113 (100.0) [64.6]	62 (100.0) [35.4]	175 (100.0)

Source: Survey Data, 2005.
Note : Figure in the parenthesis indicates column percentage.
Figure in the brackets indicates row percentage.

Public street water tanks and mobile tankers are more popular in South Chennai. Besides public water sources, households have alternative water sources also. While public water tap is the rarest but old public water distribution system

Table 5.21. Alternative Sources to Metro Water

Alternative Sources	*North Chennai*	*South Chennai*	*Total*
Own dug well	14 (21.9)	9 (7.4)	23 (12.4)
Own bore well	41 (64.1)	76 (62.8)	117 (63.2)
Packaged water	9 (14.1)	36 (29.8)	45 (24.3)
Total	64 (100.0) [34.6]	121 (100.0) [65.4]	185 (100.0)

Source: Survey Data, 2005.
Note : Figure in the parenthesis indicates column percentage.
Figure in the brackets indicates row percentage.

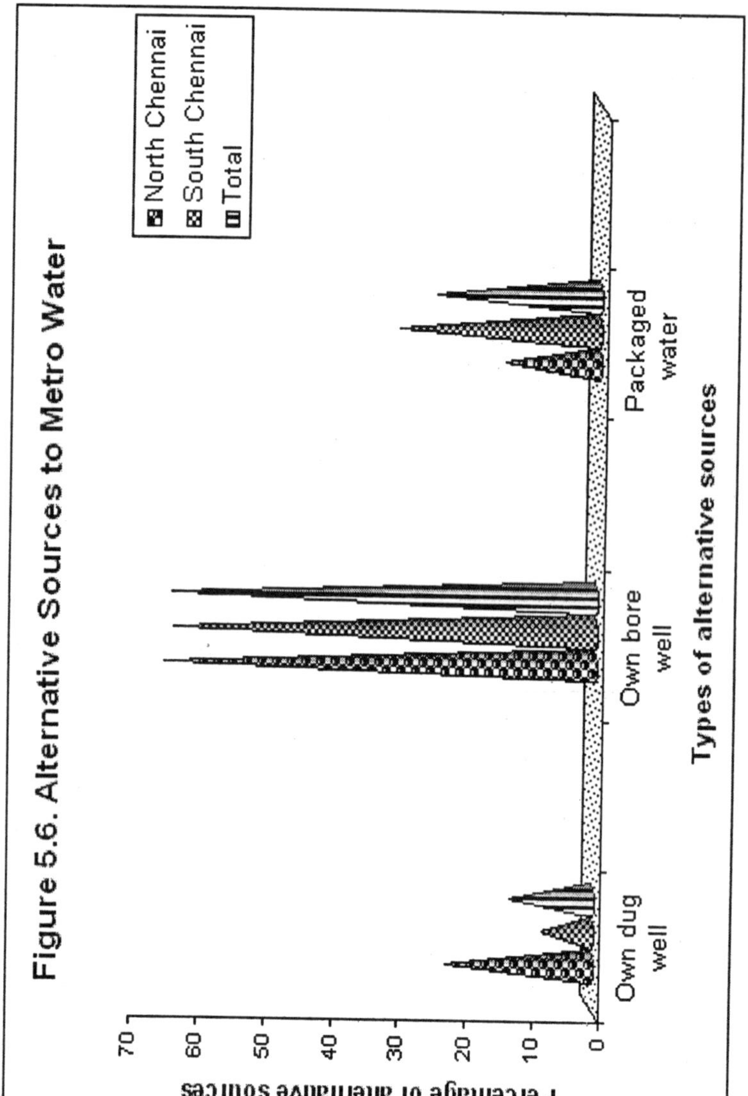

Figure 5.6. Alternative Sources to Metro Water

accounting only 3.4 per cent of the households fetching water from common source. There are 11 households (2.4 per cent) never use metro water either through in-house connection or through public water distribution point. Table 5.21 portrays the alternative water sources to metro water enjoyed by the sample households.

Three-fifths (63.2 per cent) of the households with alternative sources have bore wells. Dug- wells account for 12.4 per cent of the households having alternative water source. It is found that 24.3 per cent of the households depend on packaged water for drinking.

AVAILABILITY OF WATER BASED GADGETS OR FACILITIES

The ease and comfort of a household depends on the availability of many water based gadgets and facilities available to the households. They include dishwashers, washing machines, bath tubs, and swimming pools. The information obtained relates to the availability water equipments are furnished in Table 5.22.

With respect to the individual facilities it may be noted that 90.9 per cent of the sample households do not have dishwashers; 75.8 per cent are without washing machines, 97.8 per cent are without bath tubs and 99.8 per cent go without swimming pools. Absence of these facilities is acute in North Chennai. Considering the overall position, it can be understood that the sample households have varying extent of these facilities.

Table 5.23 gives the particulars of members taken head bath in a week.

It is found that all the households have at least one member taking bath per day. Here, the range of household

Table 5. 22. Availability of Water Equipments

Water Equipments	*Location* North Chennai		South Chennai		Total	
	Yes	*No*	*Yes*	*No*	*Yes*	*No*
Dish washer	11 (4.9)	214 (95.1)	30 (13.3)	195 (86.7)	41 (9.1)	409 (90.9)
Washing machine	33 (14.7)	192 (85.3)	76 (33.8)	149 (66.2)	109 (24.2)	341 (75.8)
Bathing tubs	4 (1.8)	221 (98.2)	6 (2.7)	219 (97.3)	10 (2.2)	440 (97.8)
Swimming pool	1 (.4)	224 (99.6)		225 (100.0)	1 (.2)	449 (99.8)

Source: Survey Data, 2005.
Note : Figure in the parenthesis indicates the percentage of total.

members varies from one to seven. What is notable is that 42.2 per cent of the households wash their clothes twice a day. Another 42.9 per cent households wash their clothes everyday.

Table 5.23. Frequency of Bath in a Week

Household members	*Location* North Chennai	*South Chennai*	*Total*
One	23 (10.2)	30 (13.3)	53 (11.8)
Two	67 (29.8)	104 (46.2)	171 (38.0)
Three	71 (31.6)	55 (24.4)	126 (28.0)
Four	37 (16.4)	28 (12.4)	65 (14.4)
Five	3 (1.3)		3 (.7)
Seven	24 (10.7)	8 (3.6)	32 (7.1)
Total	225 (100.0)	225 (100.0)	450 (100.0)

Source: Survey Data, 2005.
Note : Figure in the parenthesis indicates column percentage.

Those who possess washing machines do not wash their clothes everyday. Most of such households wash their clothes once in a week. It is found that 39.1 per cent of the households spend ten to fifteen minutes for washing laundry every day.

The frequency of laundry in a week is shown in Table 5. 24.

Table 5.24. Frequency of Laundry in a Week

Laundry takes the last week	*Location*		*Total*
	North Chennai	*South Chennai*	
Twice a day	103 (45.8)	87 (38.7)	190 (42.2)
Every day	84 (37.3)	109 (48.4)	193 (42.9)
Two to three days	24 (10.7)	26 (11.6)	50 (11.1)
Three to five days	14 (6.2)	3 (1.3)	17 (3.8)
Total	225 (100.0)	225 (100.0)	450 (100.0)

Source: Survey Data, 2005.
Note : Figure in the parenthesis indicates column percentage.

Table 5.25. Monthly Rent Statistics

Monthly Rent Statistics	*Location*		*Total*
	North Chennai	*South Chennai*	
Numbers	66	58	124
Minimum (Rs.)	250.00	350.00	250.00
Maximum (Rs.)	3500.00	7000.00	7000.00
Mean (Rs.)	987.12	1574.14	1261.70
Std. Deviation	671.93	1242.04	1019.79
Gross Rent (Rs.)	65150	91300	156450

Source: Survey Data, 2005.

The descriptive statistics of monthly rent paid by the households living in rented houses is shown in Table 5.25.

It is found that 124 of the 450 households (27.5 per cent) live in rented houses. Among them, 66 households are in North Chennai and 58 in South Chennai. The rent paid by each of the household in North Chennai ranges from Rs.250 to Rs.3500, whereas the figures for South Chennai are Rs.350 to Rs. 7000. The mean rent paid by the households in South Chennai is Rs.1574.14, compared to Rs.987.12 of the same in North Chennai. It reveals that a majority of the population lead a moderate to low level of living. The sample households in North Chennai exhibit comparatively lower socio-economic and cultural profile as compared to the households in South Chennai. Housing conditions is no exception.

6

Determinants of Residential Water Demand in Chennai City

Water distribution authorities face the problem of supplying adequate quantity and quality of water to the residential sector in urban areas. Urban water supply has become a complicated affair due to increasing scarcity of water on the one hand, and inefficient management on the other. The supply side management is fostered and nurtured by State policies and institutions; Number of residences and its population is the most significant variable determining residential water demand. Besides population, there are many other factors playing an important role in influencing the demand for domestic water supply. These factors are generally categorised into (i) economic (ii) socio-cultural (iii) environmental, and (iv) technological.

ECONOMIC FACTORS

Economic factors are more important than any other factors in the analysis of residential water demand. Here, price of water and household income are the two important economic variables. In Chennai City water is charged at a fixed rate, hence this variable obviously has no role in the residential water demand because households draw additional quantity of water at zero marginal cost. Moreover, the price of water being fixed, at an absolute amount of Rs.300 for a half-yearly period, households draw as much water as available even for non-cooking and non-drinking purposes. A majority of the households enjoy un-metered piped connection distribution system. Hence, the price or water charge does not play as an influential economic factor in deciding the residential water demand. In many studies, households income or per capita income is used as the important determinant of water demand. There is a positive relationship between income of the household's and quantity of water demanded. In other words, as the income of the household increases, its demand for water will also increase. It may be noted that the income elasticity varies between different uses of water like drinking, cooking, washing, bathing, cleaning and gardening.

SOCIO-CULTURAL FACTORS

Household size, composition of the household, level of education, average age of the members of the households, etc., are the important socio-cultural factors influencing residential water demand. Many studies (Gibbs, 1978, Foster and Beattie, 1981, Jones and Morries, 1984, Chicoine and Ramamurthy, 1986) have used the size of the household as the main socio-cultural variable. Household size influences water demand in two directions.

Increase in the size of the household would increase the total demand for water, but per capita consumption of the water may decline. This is because, a given rise in the size of the household with standard water equipments may reduce the general demand for water and the non-drinking water requirements would not increase in the same rate as does the number of persons in the household. Educational background or the educational standard of the head of the household is an important socio-cultural variable determining the demand for residential water (Darr, Fieldman and Kamen, 1975). According to them, the educational standard of the head of the household influences the domestic water demand positively. Thus, the positive impact of education on per capita consumption of water has turned out to be significant.

ENVIRONMENTAL FACTORS

Generally, the demand for water for drinking purpose will be less in high rainfall regions near the riverbanks of perennial rivers or near the well-built water reservoirs. High temperature and evapo-transpiration would enhance rather indirectly the demand for water. However, the environmental variables such as rainfall precipitation, temperature and evapo-transpiration may become important variables in cross section analysis across cities where climatic variation would be substantial in influencing the demand for water.

TECHNOLOGICAL FACTORS

Technological variables rarely find a place in the estimation of residential water demand. One of the important reasons for the exclusion of technological factors in the estimation of residential water demand is the use of secondary data, which may not provide the details of household appliances such as washing machines, dish washers, and water conserving water taps. It is also true that the type of toilet flush, showers, etc., differ widely in their water intensities. In a study, it is stated that (Postal, 1984) water conserving designs in such technologies can enhance water saving. The new technologies are not only water conserving but also cost-efficient. Therefore, technological factors play a major role in reducing water demand.

As mentioned above, there are multifarious factors influencing the demand for residential water in the metropolitan city of Chennai. Household level water access and use are discussed in the present chapter, which outlines the actual availability of water to various sections of households in Chennai City. It also tries to identify various sources of water the sample households depend upon, and the opportunity costs associated with fetching of water. Specifying the determinants of water demand, it is a difficult task to identify the most important determinants among the numerous determinants of water demand. Standard books Höglund, L. (1999) on water supply have identified some of the important determinants of urban water demand. In practical terms it is found that only a few of them have some crucial influence on water use. In this chapter a serious attempt is made to understand the effect of individual factors and their relative contribution to the variations in the household water use patterns. It may assist in taking decisions with regard to water pricing, water conservation, allocation of resource, etc. In the subsequent sections an endeavour is made to build up this knowledge by analysing all the factors one by one appearing in the demand model later.

SOURCE OF WATER SUPPLY

The only source of public water supply is the Metro Water and the nature of water supply is broadly classified into piped metro water supply and non-piped metro water supply. Piped water supplies can be classified into three kinds: i) metered in-house water connection, ii) un-metered in-house water connection, and finally, iii) in-house hand pump connection. The public water supplies provide water in four different forms: i) public tap, ii) public hand pump, iii) public street water tank and iv) mobile tanker lorry.

In-house metered water connection is based on the logic that water consumption is subject to the market economy. This will gradually wipe out the practice of treating water as free goods and in turn view it as economic goods. Usually, such water connections are for rich households consuming large quantities of water. The distribution of household income according to water sources is shown in Table 6.1.

Table 6.1. Distribution of Household Income according to Water Sources

Household Monthly Income (in Rs)	*Piped water supply*				*Type of street water collection*					*Other sources*	
	Metered	*un-metered*	*In House hand pump connection*	*Total*	*Public tap*	*Public hand pump*	*Public street water tank*	*Mobile tanker lorry*	*Total*	*Own dug well or bore well*	*Can water*
Below 2500	-	-	20 [9.3]	20 [7.6]	3 [50.0]	45 [37.5]	6 [28.6]	11 [39.3]	65 [37.1]	7 [5.3]	1 [2.2]
2500 – 5000	-	1 [2.2]	53 [24.7]	54 [20.5]	2 [33.3]	50 [41.7]	7 [33.3]	15 [53.6]	74 [42.3]	20 [15.2]	6 [13.3]
5000 – 7500	-	3 [6.5]	35 [16.3]	38 [14.4]	1 [16.7]	18 [15.0]	4 [19.0]	2 [7.1]	25 [14.3]	13 [9.8]	3 [6.7]
7500 – 10000	-	5 [10.9]	34 [15.8]	39 [14.8]	-	4 [3.3]	1 [4.8]	-	5 [2.9]	23 [17.4]	8 [17.8]

(Contd.)

(Contd. Table 6.1)

Household Monthly Income (in Rs)	Piped water supply				Type of street water collection					Other sources	
	Mete-red	un-metered	In House hand pump conne-ction	Total	Public tap	Public hand pump	Public street water tank	Mobile tanker lorry	Total	Own dug well or bore well	Can water
10000 – 12500	-	1 [2.2]	9 [4.2]	10 [3.8]	-	-	2 [9.5]	-	2 [1.1]	3 [2.3]	2 [4.4]
12500 – 15000	-	8 [17.4]	22 [10.2]	30 [11.4]	-	1 [.8]	-	-	1 [.6]	16 [12.1]	7 [15.6]
15000 and above	3 [100.0]	28 [60.9]	42 [19.5]	73 [27.7]	-	2 [1.7]	1 [4.8]	-	3 [1.7]	50 [37.9]	18 [40.0]
Total	3 [100.0]	46 [100.0]	215 [100.0]	264 [100.0]	6 [100.0]	120 [100.0]	21 [100.0]	28 [100.0]	175 [100.0]	132 [100.0]	45 [100]

Source: Survey Data, 2005.
Note : Figure in the parenthesis indicates column percentage.

It can be seen that only 1.1 per cent of the 264 households that have access to piped water supply have metered piped water connection. Over four fifths (81.4 per cent) of the households having in-house water connection get it in the form of in-house hand pump connection. Those with in-house hand pump water connection have to pump out water by exercising physical power. Hence they may not be in a position to extract water abundantly. There are 17.4 per cent of the households with piped water supply getting un-metered in-house water connection. They store the water in underground sumps and later lift it to the overhead water tank by using electrical pump sets.

Except for the metered in-house water connection, all the other categories of piped water connection Rs.300 is levied once every six months. If water is treated as economic goods, the water charge to be levied per unit volume of water is likely to influence its consumption. The water charge in the case of metered in-house connection is subject to the volume of water consumed. But the households, paying flat-water charge do not get any incentive if they use less volume of water.

There are a number of households that do not enjoy the in-house water connections. Such households fetch metro water from a common source such as public taps, public hand pumps, public street water tank and mobile tanker lorries. They account for 175 of the 450 sample households (37.7 per cent). A majority (68.6 per cent) of them depend on the public hand pump source. It shows that such sources are in the areas where water pressure is low, hence much energy is wasted for fetching water, both in terms of manual pumping of water and waiting in the queue for long hours. Moreover, it is observed that the number of *Kodam* per household is limited and if they want more water, again they have to wait in the queue at the end. The street water tanks and mobile water tanker lorries supply water to the households on a fixed quota. The street water tanks and mobile tanker lorries are the sources of metro water for 12 and 16 percentage of the households, depending on non-piped metro water supply. A majority of the slum areas in Chennai City is supplied water either by street water tanks or mobile water tanker lorries. Those areas at the tail end of the metro water distribution system and areas of low pressure are also supplied water by the means of above-cited two sources.

It is generally observed that scarcity of water often influences households to store more water than what is actually required at a time for household consumption. If the supply of water is at a price, the demand for water will significantly depend on its price. The real cost of fetching water is one of the important factors that affects the residential water demand and an increased water price will reduce the consumption of water in metered areas. In Chennai City, provision of water is a sensitive and more populist issue because of the democratic polity; hence the authorities cite the reason that metering of water is an uphill task. They often cite some reasons like meters go non-operational and additional manpower is required for taking household water consumption accounts in their attempt to avoid progressive water pricing policy.

It is revealed in the survey that 132 of the 450 (29.5 per cent) households use their own dug-wells or bore-wells for supplementing the water requirements. There are 45 of the 450 households (10 per cent) that buy packaged water for the purpose of drinking and cooking.

The general economic theory explains that there is a cause and effect relationship between consumer's income and the sources and level of consumption of water. Since water charge is flat for a majority of the households with in-house water connection and the households getting water from street sources do so completely free of cost, it is interesting to compare the sources of water and the levels of household income (see Table 6.1). All the households connected to the metered in-house piped connection belong to the highest monthly income slab of Rs.15000 and above. It is also revealed that the percentage of households connected to the un-metered in-house piped water connection increases as the level of household income increases. Three-fifths (60 per cent) of such households are in the highest household monthly income slab. The in-house hand pump connection does not show any clear association with the level of household incomes.

When an analysis of the non-piped water sources is made, it is found that all the households using public taps and mobile tanker lorries are represented in the lower income groups. A majority of the households that use public hand pumps and the public street water tanks are concentrated in the areas of lower income groups. It is also learnt that nearly two-fifths (37.9 per cent) of the households belonging to the highest monthly household income group of Rs.15000 and above possess own dug wells or tube-wells. This is also true in the case of the households buying packaged drinking water.

If there is flat price, the demand for a commodity will depend upon the income of the household or the individual. But empirical evidence establishes that this relationship will not hold in the case of water. The income of the household is an indicator of the ability to pay for the services. Higher income households live in better housing conditions and keep certain water equipments for increasing their water convenience. Usually they install more water based appliances like showers, water taps, luxurious toilet devices, washing machines, dishwashers, Jacuzzis etc. It is generally accepted that the per capita cooking and drinking demand for water is more or less the same for households in different income groups. However, water demand for other uses varies across households with different incomes. For such uses, the relationship between the income of a household and consumption of water can be expressed in terms of income elasticity. In developed countries, where water consumption is metered and priced, they exhibit a positive correlation between per capita residential use of water and household income (Larson and Hudson, 1951). In Chennai City, the household demand for water is determined by various economic, social, cultural and demographic characteristics of the household. In addition to this, the household demand for water is also influenced by metro water availability, alternative sources of water, and price of private water sources.

MONTHLY HOUSEHOLD INCOME AND LITRES PER CAPITA CONSUMPTION PER DAY (LPCD)

Table 6.2 gives the distribution of water consumption according to monthly household income. It can be seen that the per capita consumption of metro water differs among different income groups. A general trend that emerges is that higher the level of household income, lesser will be the per capita metro water consumption per day.

This indicates that the access to metro water by the poor living in slums is more because they expect metro water to supply almost all-household water needs since alternatives are not available before them. They solely depend on metro water, as they can not afford to buy water from private sources. However, the higher income people purchase water for drinking and other household needs from private sources in order to beat inadequate metro water supply. Hence the per capita metro water consumption per day by the higher income households appears slightly lower than that of the lower income households.

The litre per capita consumption per day (lpcd) of metro water in different locations of Chennai City is shown in Table 6.3. It may be noted that 11 of the 450 households (2.4 per cent) do not depend on metro water for their household water needs. Such households are 4 and 7 in North and South Chennai, respectively. Among those households depending on metro water, more than 40 per cent belong to the category of 60 to 80

Table 6.2. Distribution of Water Consumption according to the Monthly Household Income

Monthly Household Income (in Rs.)	*Sample household members*	*CMWSSB water supply*		
		Monthly gross Water consumption (in liters)	*Daily gross Water Consumption (in liters)*	*Average per capita water consumption per day (in Liters)*
Below 2500	176	420300	14010	79.60
2500 - 5000	283	686250	22875	80.83
5000 - 7500	146	359100	11970	81.99
7500 - 10000	77	197550	6585	67.88
10000 - 12500	28	62100	2070	73.93
12500 - 15000	45	133650	4455	68.54
15000 and above	166	284850	9495	57.25
Total	961	2143800	71460	74.36

Source: Survey Data, 2005.

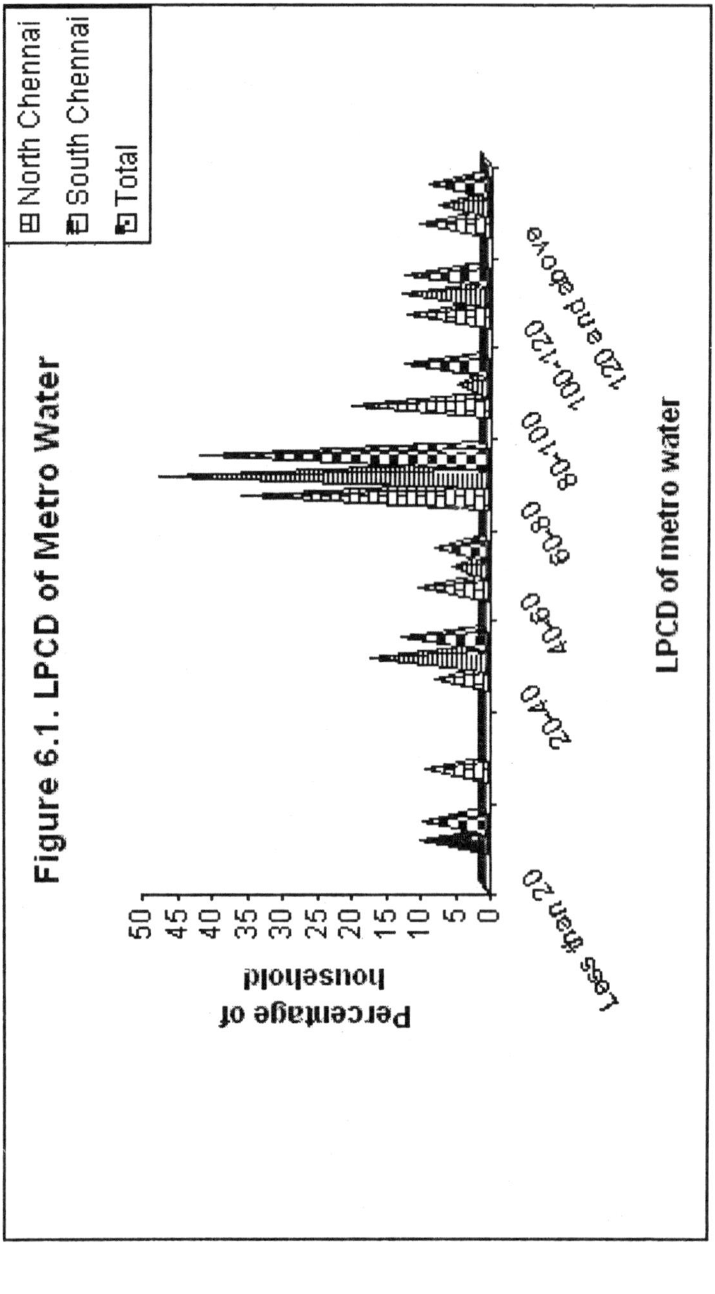

Figure 6.1. LPCD of Metro Water

litres per capita consumption per day. The percentage of households, consuming more than 60 to 80 lpcd in North Chennai is 39.8 as compared to the 22.9 per cent of the same in South Chennai. Moreover in South Chennai, the percentage share of households in lower lpcd levels is less as compared to the percentage share of households in North Chennai.

It is inferred that households in South Chennai consume less lpcd of water as compared to the households in North Chennai. It may also be due to the fact that the practice of marketisation of water in North Chennai appears less as compared to that of the same in South Chennai.

Table 6.3. Liters Per Capita Consumption Per Day of Metro Water

Litres per capita consumption per day	*Location*		*Total*
	North Chennai	*South Chennai*	
Less than 20	19 [8.6]	20 [9.2]	39 [8.9]
20-40	16 [7.2]	36 [16.5]	52 [11.8]
40-60	21 [9.5]	10 [4.6]	31 [7.1]
60-80	77 [34.8]	102 [46.8]	179 [40.8]
80-100	42 [19.0]	9 [4.1]	51 [11.6]
100-120	25 [11.3]	26 [11.9]	51 [11.6]
120 and above	21 [9.5]	15 [6.9]	36 [8.2]
Total	221 [100.0]	218 [100.0]	439 [100.0]

Source: Survey Data, 2005.
Note : Figure in the brackets indicates column percentage.

LITERACY STANDARD AND LPCD OF CMWSSB WATER

The literacy standard of an individual largely influences his hygienic sense. Higher standard of educational levels

Table 6.4. LPCD according to Educational Standard of Heads of the Households

Lpcd	*Educational Standard of Head of households*									
	Illi-terate	*Pri-mary*	*Middle*	*High school*	*Higher Second-dary / PUC*	*Gra-duate*	*Post-grad-uate*	*Prof-essional*	*Others*	*Total*
Less than 20	4 (12.9) [4.4]		3 (9.7) [3.9]	7 (22.6) [5.2]	2 (6.5) [5.0]	8 (25.8) [17.4]	4 (12.9) [20.0]	3 (9.7) [60.0]		31 (100.0) [7.2]
20-40	3 (5.8) [3.3]		9 (17.3) [11.8]	10 (19.2) [7.5]	9 (17.3) [22.5]	11 (21.2) [23.9]	8 (15.4) [40.0]	2 (3.8) [40.0]		52 (100.0) [12.1]
40-60	6 (19.4) [6.6]	4 (12.9) [22.2]	4 (12.9) [5.3]	7 (22.6) [5.2]	7 (22.6) [17.5]	2 (6.5) [4.3]			1 (3.2) [100.0]	31 (100.0) [7.2]
60-80	49 (27.4) [53.8]	10 (5.6) [55.6]	33 (18.4) [43.4]	61 (34.1) [45.5]	9 (5.0) [22.5]	14 (7.8) [30.4]	3 (1.7) [15.0]			179 (100.0) [41.5]

(Contd.)

(Contd. Table 6.4)

Lpcd	*Educational Standard of Head of households*									
	Illi-terate	*Pri-mary*	*Middle*	*High school*	*Higher Second-dary / PUC*	*Gra-duate*	*Post-grad-uate*	*Prof-essional*	*Others*	*Total*
80-100	15 (29.4) [16.5]	2 (3.9) [11.1]	11 (21.6) [14.5]	16 (31.4) [11.9]	3 (5.9) [7.5]	4 (7.8) [8.7]				51 (100.0) [11.8]
100-120	6 (11.8) [6.6]	1 (2.0) [5.6]	8 (15.7) [10.5]	20 (39.2) [14.9]	6 (11.8) [15.0]	6 (11.8) [13.0]	4 (7.8) [20.0]			51 (100.0) [11.8]
120 and above	8 (22.2) [8.8]	1 (2.8) [5.6]	8 (22.2) [10.5]	13 (36.1) [9.7]	4 (11.1) [10.0]	1 (2.8) [2.2]	1 (2.8) [5.0]			36 (100.0) [8.4]
Total	91 (21.1) [100.0]	18 (4.2) [100.0]	76 (17.6) [100.0]	134 (31.1) [100.0]	40 (9.3) [100.0]	46 (10.7) [100.0]	20 (4.6) [100.0]	5 (1.2) [100.0]	1 (.2) [100.0]	431 (100.0) [100.0]

Source: Survey Data, 2005.
Note : Figure in the parenthesis indicates row percentage.
Figure in the brackets indicates column percentage.

persuade people to follow various hygiene habits requiring more quantity of water for consumption. People who lack education may cause unusually low consumption of water, which increases the risk of water-related diseases. Logically there is a positive degree of relationship between the level of education and water use. In order to compare the relationship between educational levels and the quantum of household water consumption, the educational standards of the head of the sample households are compared with the use of lpcd of water from CMWSSB. The results are furnished in Table 6.4.

It is found that a majority of the households headed by illiterates and educated up to high school level consume 60-80 lpcd of metro water. The number of households that consume less quantity than this consumption level of metro water is found less in number when compared to the number of households above this consumption level. However, households with heads

Table 6.5. Source of Water Consumption According to the Monthly Household Income

Household monthly Income (Rs.)	*Average per capita Metro water consumption per day*	*Average per capita total water consumption per day*	*% share of metro water in total water consumption*
Below 2500	79.60	109.77	69.8
2500 - 5000	80.83	125.67	64.3
5000 - 7500	81.99	127.71	64.2
7500 - 10000	67.88	135.56	50.1
10000 - 12500	73.93	135.00	54.8
12500 - 15000	68.54	165.46	41.4
15000 and above	57.25	194.82	29.4
Total	74.36	142.18	52.3

Source: Survey Data, 2005.
Note : 1.Metro water consumption includes water supplied by CMWSSB, 2.Total water consumption includes water supplied by CMWSSB, own sources and private water supply.

educated up to graduate level and above are found in the lower levels of consumption of metro water. It may be due to the fact that the better educated heads belong to the households with better economic status.

Households with better economic status depend more on reliable private water market and also avail whatever quantity from metro water. In order to assess the quantity of water consumption of the poor and the well-to-do, it has been decided to compare the quantity of metro water consumption and total water consumption of the sample households. Table 6.5 shows the average metro water consumption and total water consumption of households in different income levels.

It is interesting to see that at lower levels of monthly household income higher is the average per capita consumption of CMWSSB water (metro water). While at higher levels of monthly household income, higher is the average per capita total water but lesser the average per capita metro water consumption. It is because that the households at higher levels of income are not prepared to take a higher unreliability cost due to the negative externalities of metro water. Hence they opt for private water sources, even though the unit cost of water of the private source is much higher than that of the public sources. A comparison is made by taking the percentage share of metro water consumption to total water consumption at different household income levels.

The pattern clearly emerges that the households at lower income levels depend more on public sources metro water, while households at the higher income levels showed a greater dependence on private water sources to meet not only their primary uses such as drinking and cooking but also for secondary uses like bathing and cleaning. The quality of water of the public source is the same at the point of delivery for all purposes. However, the private water quality ranges from the packaged drinking water to shallow water for bathing and cleaning.

Figure.6.2. Source of Water Consumption and Monthly Household Income

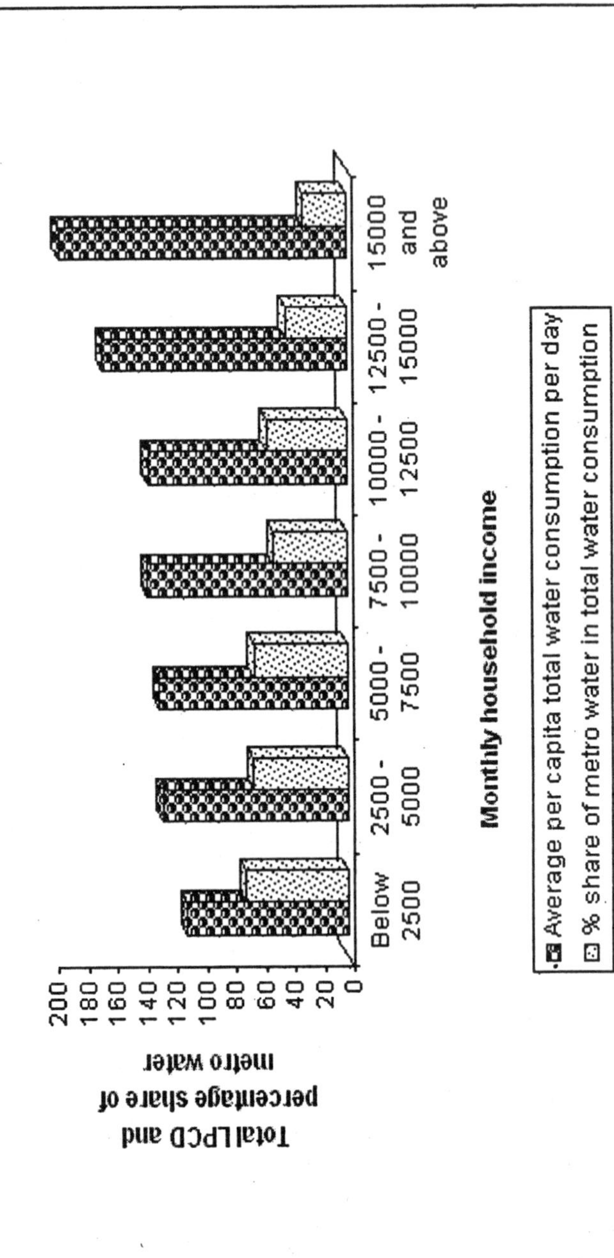

Table 6.6. Distribution of Monthly Household Income according to Litres per Capita Consumption per Day

Litres per capita consumption per day	*Monthly Household Income (in Rs.)*							*Total*
	Below 2500	*2500 - 5000*	*5000 - 7500*	*7500 - 10000*	*10000 - 12500*	*12500 - 15000*	*15000 and above*	
Less than 20	1 [1.2]	6 [4.8]	4 [6.3]	3 [7.0]	2 [16.7]	3 [10.0]	12 [16.0]	31 [7.2]
20-40	3 [3.6]	3 [2.4]	2 [3.2]	7 [16.3]	1 [8.3]	6 [20.0]	30 [40.0]	52 [12.1]
40-60	9 [10.8]	10 [8.0]	5 [7.9]	3 [7.0]	-	-	4 [5.3]	31 [7.2]
60-80	42 [50.6]	61 [48.8]	28 [44.4]	20 [46.5]	5 [41.7]	13 [43.3]	10 [13.3]	179 [41.5]
80-100	16 [19.3]	16 [12.8]	10 [15.9]	5 [11.6]	1 [8.3]	2 [6.7]	1 [1.3]	51 [11.8]
100-120	5 [6.0]	16 [12.8]	9 [14.3]	4 [9.3]	2 [16.7]	4 [13.3]	11 [14.7]	51 [11.8]
120 and above	7 [8.4]	13 [10.4]	5 [7.9]	1 [2.3]	1 [8.3]	2 [6.7]	7 [9.3]	36 [8.4]
Total	83 [100.0]	125 [100.0]	63 [100.0]	43 [100.0]	12 [100.0]	30 [100.0]	75 [100.0]	431 [100.0]

Source: Survey Data, 2005.

Note : Figure in the brackets indicates column percentage.

Per Capita Public Water Consumption And Household Income

In order to understand the impact of household income on the per capita metro water consumption, various levels of water consumption have been compared with various levels of household incomes. The results are furnished in Table 6.6.

Though water is priced on a flat rate to the households with in-house un-metered water connections, progressive rates for the metered consumption, and free water is supplied through common sources, the level of consumption is not strictly based on the level of household income. It is found that two-fifths (40 per cent) of the households in the highest monthly income slab of Rs.15, 000 and more availed public water to the levels of 20 to 40 lpcd. However, more than half of the households in the lowest monthly household income level of below Rs.2500 consumed public water to the tune of 60 – 80 lpcd.

More than one-fourth (29 per cent) of the households are located in the monthly income slab of Rs.2500 to Rs.5000. Out of these 84.4 per cent consumed public water to the level of 60 lpcd or more. This pattern of water consumption is visible at all the levels of income. While viewing the lpcd levels of public water at various household income levels, it is learnt that at the lowest lpcd level of less than 20 litres, more than half (54.9 per cent) of the households are rich households (monthly household income levels of Rs.10000 or more). It is 71.1 per cent in the case of 20-40 lpcd levels. Thereafter, when the levels of public water consumption increase, more households are found in the lower income groups. It is therefore clear that the households at the higher income levels depend less on public water sources, though their level of water consumption is much higher than those of the households at the lower income levels.

Public Water Consumption and The Size of The Household

The household size has a direct bearing on the total household consumption of water. However, it is necessary to compare the household size with the total quantum of consumption of public water. Generally any increase or decrease in household size causes an increase or decrease in the consumption of water by the household. But there is an economy of scale in water use that an increase in family size reduces the rate of increase in total consumption of water. The causes of the economies of scale of water use are such as food preparation, general washing, cleaning, and watering the plants, car washing and similar activities where there is a sharing of consumption of water. According to Howe and Linaweaver (1967), the household size strongly influences the daily domestic demand for water. The study, however, does not provide anything regarding the economies of scale and this factor has been different in different areas. Probably the variation in family size with metered use of water may be different from the un-metered use of water based on flat rate pricing. Again, the total household water demand may be based on the distance of water source from the house, if outdoor water is to be procured. Based on these views, in Table 6.7, the household size has been compared with the per capita consumption of public water (lpcd).

While analysing Table 6.7 it becomes evident that a majority of the households in the household size of below 2, 2-4, and 4-6 members, have the percentages of 47.8, 50.7 and 34.5 per cent respectively, and consume 60 - 80 lpcd of public water. When the household's size further increases, the lpcd of water also increases. For instance, the maximum number (38.5 per cent) of households in the family size of 6-8 falls in the water consumption category of 80-100 lpcd. Moreover, 75 per cent of the households having a household size of 8 and above consume 120 and more lpcd. Then, it goes against the contention of the concept of economies of scale whereby the per capita consumption of water declines when the household size increases. This is mainly due to the fact that the consumption level of water for drinking and cooking may not increase proportionately

Table 6. 7. Distribution of Litres per Capita Consumption Per Day according to the Household Size

Litres per capita consumption per day	*Size of household (in Numbers)*					*Total*
	Below 2	*2 - 4*	*4 - 6*	*6 - 8*	*8 and above*	
Less than 20	8	12	5	5	1	31
	[17.4]	[5.8]	[3.4]	[19.2]	[25.0]	[7.2]
20-40	4	32	14	2		52
	[8.7]	[15.5]	[9.5]	[7.7]		[12.1]
40-60	8	14	8	1		31
	[17.4]	[6.8]	[5.4]	[3.8]		[7.2]
60-80	22	105	51	1		179
	[47.8]	[50.7]	[34.5]	[3.8]		[41.5]
80-100	2	16	23	10		51
	[4.3]	[7.7]	[15.5]	[38.5]		[11.8]
100-120	1	22	27	1		51
	[2.2]	[10.6]	[18.2]	[3.8]		[11.8]
120 and above	1	6	20	6	3	36
	[2.2]	[2.9]	[13.5]	[23.1]	[75.0]	[8.4]
Total	46	207	148	26	4	431
	[100.0]	[100.0]	[100.0]	[100.0]	[100.0]	[100.0]

Source: Survey Data, 2005.
Note : Figure in the brackets indicates column percentage.

with increases in the household's size. However, it increases for the purposes of washing and bathing.

Water Consumption and Water-Using Appliances

Water consumption depends on the number and type of water-using appliances possessed by a household. More number of appliances will induce more water consumption. A survey of availability of water using appliances by the households therefore is useful and in many cases it will provide justification for variations in the consumption of water amongst the households with similar characteristics. In the views of March (1972), a change in water consumption is associated with a change in the number and type of plumbing fixtures. According to White, et. al., (1972), the number of plumbing fixtures is an index of the

Table 6.8. Distribution of Litres per Capita Consumption per Day according to the Number of Bath Showers

Litres per capita consumption per day	*Number of showers*		*Total Use (Liters)*
	One	*Two*	
Less than 20	17 [16.0]	3 [60.0]	20 [18.0]
20-40	34 [32.1]	1 [20.0]	35 [31.5]
40-60	5 [4.7]		5 [4.5]
60-80	24 [22.6]		24 [21.6]
80-100	2 [1.9]	1 [20.0]	3 [2.7]
100-120	17 [16.0]		17 [15.3]
120 and above	7 [6.6]		7 [6.3]
Total	106 [100.0]	5 [100.0]	111 [100.0]

Source: Survey Data, 2005.
Note : Figure in the brackets indicates column percentage.

extent of water consumption. The details of distribution of liters per capita consumption per day according to the number of bath showers are shown in Table 6.8.

It is found that 111 of the 450 sample households (24.6 per cent) use showers in their houses. Among the total households with bath showers, 95.5 per cent have single showers. It is interesting to see that households with 2 showers are in the category of consuming less public water, i.e., 80 per cent of such households consume less the 40 lpcd.

Table 6.9 shows the distribution of liters per capita consumption per day according to the number of flushing toilets. It is evident that 408 of the 450 sample households (90.6 per cent) have flushing toilets in their houses. The number of flushing toilets varies from household to household. Households with one, two three and four flushing toilets are in the percentages of 69.4, 29.21, 2.0 and 0.5 respectively. More than two fifths (44.6 per cent) of the households with single flushing toilets consume 60-80 lpcd; 29.8 per cent of the households having two flushing toilets consume 60-80 lpcd.

However, 50 per cent of the households having three flushing toilets consume 40-60 lpcd only. The households with four flushing toilets are only 3 in number who also consume less than 80 lpcd. Some of the conventional water-flushed toilets cause a reduction in daily per person water use for flushing. Distribution of litres per capita consumption per day according to the number of the kitchen sinks used by the sample households is shown in Table 6.10. Out of 450 sample households, 209 (46.4 per cent) have kitchen sinks in their households. The number of kitchen sinks also varies from one to four. Of those households with kitchen sinks, 145 (69.4 per cent) have a single sink each in their kitchen. Households with two, three and four kitchen sinks are in the percentages of 29.2,1 and 0.5 respectively. Most of the households with more kitchen sinks consume less public water.

Table 6.9. Distribution of LPCD according to the Number of Flushing Toilets

Per capita average metro water consumption per day	*Number of flushing toilets*				*Total*
	One	*Two*	*Three*	*Four*	
Less than 20	13 [4.7]	17 [14.0]	1 [12.5]	-	31[7.6]
20-40	20 [7.2]	27 [22.3]	4 [50.0]	1 [33.3]	52[12.7]
40-60	23[8.3]	7[5.8]	-	1 [33.3]	31 [7.6]
60-80	123 [44.6]	36 [29.8]	3 [37.5]	1 [33.3]	163[40.0]
80-100	37 [13.4]	10 [8.3]	-	-	47 [11.5]
100-120	35 [12.7]	15[12.4]	-	-	50 [12.3]
120 and above	25 [9.1]	9 [7.4]	-	-	34 [8.3]
Total	276[100.0]	121[100.0]	8 [100.0]	3 [100.0]	408[100.0]

Source: Survey Data, 2005.

Note : Figure in the brackets indicates column percentage.

Table 6.10. Distribution of Litres per Capita Consumption Per Day according to the Number of Kitchen Sink

Litres per capita consumption per day	*Number of kitchen sinks*				*Total*
	One	*Two*	*Three*	*Four*	
Less than 20	16	11	-	-	27
	[11.0]	[18.0]			[12.9]
20-40	31	16	2	-	49
	[21.4]	[26.2]	[100.0]		[23.4]
40-60	10	3	-	-	13
	[6.9]	[4.9]			[6.2]
60-80	47	18	-	1	66
	[32.4]	[29.5]		[100.0]	[31.6]
80-100	8	1	-	-	9
	[5.5]	[1.6]			[4.3]
100-120	23	8	-	-	31
	[15.9]	[13.1]			[14.8]
120 and above	10	4	-	-	14
	[6.9]	[6.6]			[6.7]
Total	145	61	2	1	209
	[100.0]	[100.0]	[100.0]	[100.0]	[100.0]

Source: Survey Data, 2005.
Note : Figure in the brackets indicates column percentage.

Table 6. 11. Distribution of Litres Per Capita Consumption Per Day according to the Number of Bath Rooms

Litres per capita consumption per day	*Number of bath rooms*				*Total*
	One	*Two*	*Three*	*Four*	
Less than 20	13 [4.5]	17 [14.5]	1 [8.3]	-	31[7.4]
20-40	19 [6.6]	28 [23.9]	4 [33.3]	1 [33.3]	52 [12.4]
40-60	25 [8.7]	5[4.3]	-	1 [33.3]	31[7.4]
60-80	130 [45.3]	35 [29.9]	5 [41.7]	1 [33.3]	171[40.8]
80-100	40 [13.9]	8 [6.8]	1[8.3]	-	49 [11.7]
100-120	34 [11.8]	15 [12.8]	1 [8.3]	-	50[11.9]
120 and above	26 [9.1]	9 [7.7]	-	-	35 [8.4]
Total	287[100.0]	117 [100.0]	12 [100.0]	3 [100.0]	419[100.0]

Source: Survey Data, 2005.
Note : Figure in the brackets indicates column percentage.

Table 6.12. Distribution of Litres Per Capita Consumption Per Day according to the number of wash Basins

Litres per capita consumption per day	*Number of wash basins*			*Total*
	One	*Two*	*Three*	
Less than 20	13 [10.6]	11 [17.5]	-	24 [12.7]
20-40	27 [22.0]	16 [25.4]	3 [100.0]	46 [24.3]
40-60	10 [8.1]	1 [1.6]	-	11 [5.8]
60-80	38 [30.9]	21 [33.3]	-	59 [31.2]
80-100	6 [4.9]	3 [4.8]	-	9 [4.8]
100-120	22 [17.9]	9 [14.3]	-	31 [16.4]
120 and above	7 [5.7]	2 [3.2]	-	9 [4.8]
Total	123 [100.0]	63 [100.0]	3 [100.0]	189 [100.0]

Source: Survey Data, 2005.

Note : Figure in the brackets indicates column percentage.

It is not quite true that households with more number of kitchen sinks consume more water. A maximum number of households using one and two kitchen sinks consume 60-80 lpcd, and they are in the percentages of 32.4 and 29.5 respectively.

The number of bathrooms used by the households is taken as one of the indices of water consumption. Table 6.11 gives the distribution of litres per capita consumption per day of public water according to the number of bathrooms.

Out of the 450 sample households, 419 possess at least one bathroom. It varies from one to four in the percentages of 68.5, 27.9 and 0.7 respectively. Maximum number of households with a bathroom (45.3 per cent) consumes 60-80 lpcd of water. But when the number of bathrooms increases, not more households are found in the higher lpcd using ranges.

The distribution of litres per capita consumption per day according to the number of washbasins used is shown in the Table 6.12. It is clear that 189 of the 450 households (42 per cent) possessed washbasins in their households.

It is revealed that only 1. 6 per cent of the households have three washbasins and when the number of wash basins increase, the number of households decreases. A majority of the households with washbasins (65.1 per cent) possess single one. It is interesting to note that households with more washbasins are found in the lower classes of public water consumption.

Table 6.13 gives the details of distribution of litres per capita consumption per day according to the availability of dish washer, washing machine and bathing tubs. A general trend emerges that most of the households with washing machines are found in the lower metro water consumption classes. It is evident that only 8.8 per cent of the sample households possess

Table 6.13. Distribution of Litres Per Capita Consumption Per Day according to the Availability of Dish washer, Washing Machine and Bathing Tubs

Litres per capita consumption per day	*Number of wash basins*			*Total*
	One	*Two*	*Three*	
Less than 20	13 [10.6]	11 [17.5]	-	24 [12.7]
20-40	27 [22.0]	16 [25.4]	3 [100.0]	46 [24.3]
40-60	10 [8.1]	1 [1.6]	-	11 [5.8]
60-80	38 [30.9]	21 [33.3]	-	59 [31.2]
80-100	6 [4.9]	3 [4.8]	-	9 [4.8]
100-120	22 [17.9]	9 [14.3]	-	31 [16.4]
120 and above	7 [5.7]	2 [3.2]	-	9 [4.8]
Total	123 [100.0]	63 [100.0]	3 [100.0]	189 [100.0]

Source: Survey Data, 2005.
Note : Figure in the brackets indicates column percentage.

dishwashers. But it is interesting to note that most of the dishwasher-using households are also found in the lower levels of metro water consumption.

It is interesting to note that most of the rich household possesses dish washers and they depend less on metro water. The conclusion drawn from dishwasher usage is equally applicable to the households using washing machines and bathing tubs. Generally the rich households who consume more quantity of water operate the water using appliances. The present study has inferred that such households depend less on metro water and they may be using some alternative sources of water either from own sources or from the market for all these purposes. It is generally perceived that the floor area of the house is

Table 6. 14. Distribution of Litres per Capita Consumption per Day according to the Area of the House

Litres per capita consumption per day	*Total floor area of the house (in sq.ft.)*						
	Less than 300	*300-600*	*600-900*	*900-1200*	*1200-1500*	*1500 and above*	*Total*
Less than 20	-	2 [1.4]	6 [5.2]	11 [15.1]	8 [14.3]	4 [14.8]	31 [7.2]
20-40	-	3 [2.2]	13 [11.2]	12 [16.4]	14 [25.0]	10 [37.0]	52 [12.1]
40-60	1 [5.0]	10 [7.2]	9 [7.8]	8 [11.0]	2 [3.6]	1 [3.7]	31 [7.2]
60-80	11 [55.0]	77 [55.4]	49 [42.2]	21 [28.8]	12 [21.4]	9 [33.3]	179 [41.5]
80-100	4 [20.0]	26 [18.7]	8 [6.9]	7 [9.6]	6 [10.7]	-	51 [11.8]
100-120	3 [15.0]	10 [7.2]	21 [18.1]	8 [11.0]	8 [14.3]	1 [3.7]	51 [11.8]
120 and above	1 [5.0]	11 [7.9]	10 [8.6]	6 [8.2]	6 10.7]	2 [7.4]	36 [8.4]
Total	20 [100.0]	139 [100.0]	116 [100.0]	73 [100.0]	56 [100.0]	27 [100.0]	431 [100.0]

Source: Survey Data, 2005.
Note : Figure in the brackets indicates column percentage.

positively related to extent of water consumption. Hence, the larger the house more will be consumption of water.

Table.6.14 shows the distribution of Liters per capita consumption per day according to the area of the house.

The households representing the smallest floor area of less than 300 sq.ft accounts for 4.6 per cent. No household from this category represent the lowest per capita average metro water consumption per day of less than 20 litres and 20-40 litres. It could be seen that the representation of households in the larger floor space are found in the classes of lower metro water consumption, which is a contradictory outcome against the common perception. The slum households generally collect water from only one or two taps or steel tanks, which obviously results in a lower per-person consumption of water. It is found that more than two-fifths of the households (41.5 per cent) consume 60-80 litres per day per capita of water, of which 43 per cent lived in houses having an area of 300 to 600 sq.ft.

Households belonging to the lower levels of water consumption are generally found in bigger houses and a majority of the households with higher public water consumption are found relatively in the houses of lesser floor area. From the above analysis, it is quite clear that the poor households depend more on metro water as compared to the rich people.

TOTAL WATER SUPPLY

The household water consumption basket constitutes metro water plus private water. Private water is derived both from own source and from the water market. Dug-wells and bore-wells are the private water sources, involving some installation costs at the beginning, followed by operation and maintenance costs. Some slum households tap water from the public shallow bore-wells free of cost, where the users share the operation and maintenance costs. Many of the households in the upper middle and higher income groups purchase water from the market both for drinking and other purposes. For drinking and cooking, households buy packaged water whereas for other purposes they buy tanker lorry water and store it in underground tanks. Thus the water basket constitutes a water mix of both public and private water.

In order to examine the differences in the water consumption pattern of households, the total per capita per day water consumption is compared with various social and economic factors. The distribution of the size of the households according to the total per capita per day consumption of water is shown in Table 6.15. The per capita per day water consumption of the sample households in Chennai City show that nearly half (47.1 per cent) of them consume 50 to 75 litres per capita per day. Only 4.4 per cent of households consume less than 50 litres per capita per day. However, the number of households consuming more than 75 litres per day decreases when the water consumption level increases. It is evident that the per capita per day consumption of water does not make any significant change, when the household size increases. Households with a size of less than 2 members and 2 to 4 members are concentrated in the total water consumption levels of 50 to 75 lpcd in the percentages of 54 and 58.9 respectively.

However, it is found that 80.5 per cent of the households with a family size of 4 to 6 members consume water between 50 lpcd and 125 lpcd. The general tendency is that when the size of household increases, the lpcd of total water also increases.

Table 6.16 shows the distribution of the per capita per day consumption of water according to the monthly household income.

It is quite clear that the maximum number of households in all the income classes except the highest income class is

Table 6. 15. Distribution of Total Per Capita Per Day Consumption of Water according to the Household Size

Total Per Capita Per Day Consumption of Water (in litres)	*Size of household*					*Total*
	Below 2	*2 - 4*	*4 - 6*	*6 - 8*	*8 and above*	
Below 25	-	-	1 [0.6]	-	-	1 [0.2]
25-50	2 [4.0]	7 [3.3]	7 [4.5]	3 [10.7]	-	19 [4.2]
50-75	27 [54.0]	126 [58.9]	53 [34.4]	5 [17.9]	1 [25.0]	212 [47.1]
75-100	8 [16.0]	39 [18.2]	34 [22.1]	11 [39.3]	-	92 [20.4]
100-125	9 [18.0]	30 [14.0]	37 [24.0]	6 [21.4]	2 [50.0]	84 [18.7]
125-150	1 [2.0]	7 [3.3]	14 [9.1]	2 [7.1]	1 [25.0]	25 [5.6]
150 and above	3 [6.0]	5 [2.3]	8 [5.2]	1 [3.6]	-	17 [3.8]
Total	50 [100.0]	214 [100.0]	154 [100.0]	28 [100.0]	4 [100.0]	450 [100.0]

Source: Survey Data, 2005.
Note : Figure in the brackets indicates column percentage.

Figure 6.3. LPCD of Metro Water and Household Size

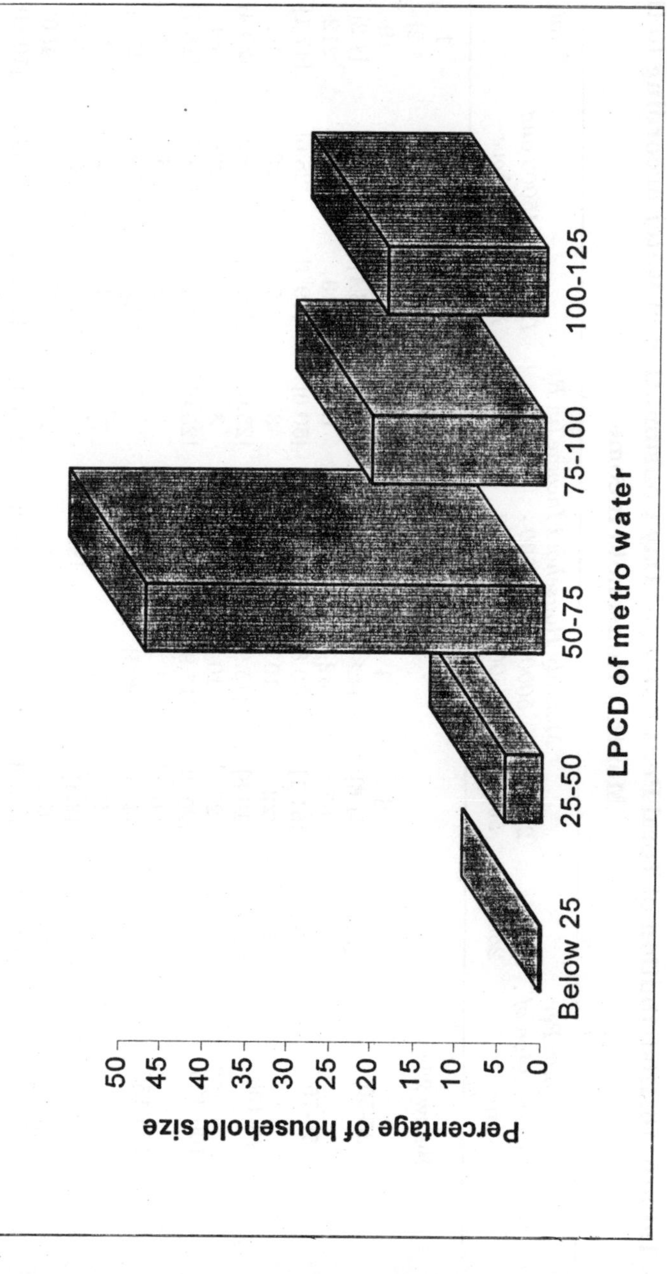

Table 6. 16. Distribution of Total Per Capita per Day Consumption of Water according to the Monthly Household Income

Total Per Capita Per Day Consumpt-ion of Water	*Monthly Household Income (in Rs.)*							*Total*
	Below 2500	*2500 - 5000*	*5000 - 7500*	*7500 - 10000*	*10000 - 12500*	*12500 - 15000*	*15000 and above*	
Below 25	1 [1.1]	-	-	-	-	-	-	1 [.2]
25-50	3 [3.4]	6 [4.6]	3 [4.8]	3 [6.4]	-	2 [6.1]	2 [2.6]	19 [4.2]
50-75	51 [58.6]	67 [51.1]	32 [50.8]	21 [44.7]	6 [50.0]	13 [39.4]	22 [28.6]	212 [47.1]
75-100	19 [21.8]	22 [16.8]	10 [15.9]	8 [17.0]	3 [25.0]	7 [21.2]	23 [29.9]	92 [20.4]
100-125	6 [6.9]	27 [20.6]	10 [15.9]	9 [19.1]	3 [25.0]	7 [21.2]	22 [28.6]	84 [18.7]
125-150	4 [4.6]	6 [4.6]	4 [6.3]	2 [4.3]	-	3 [9.1]	6 [7.8]	25 [5.6]
150 and above	3 [3.4]	3 [2.3]	4 [6.3]	4 [8.5]	-	1 [3.0]	2 [2.6]	17 [3.8]
Total	87 [100.0]	131 [100.0]	63 [100.0]	47 [100.0]	12 [100.0]	33 [100.0]	77 [100.0]	450 [100.0]

Source: Survey Data, 2005.
Note : Figure in the brackets indicates column percentage.

found in the per capita per day water consumption level of 50-75 litres. However the percentage share of households in this water consumption class decreases when the income level increases. The general trend is that higher percentage share of households in the higher income levels are found in the higher classes of per capita per day water consumption levels. Hence it is clear that there exists a positive relationship between total water consumption and monthly household incomes. But this relationship could not be established in the case of metro water.

In order to assess the consumption of water by the households connected with in-house metro water and those that depend on public sources, the per capita per day consumption of water has been compared with the sources of water. The results are furnished in Table 6.17.

It is found that more than three-fifths (60.1 per cent) of the households enjoy private user metro water connection within their residence. However, 39.9 per cent of the households depend on the common sources of water outside their residences. The inference derived here is that the households with in-house

Table 6.17. Distribution of Sources of Water according to LCPD Consumption of Water

LCPD	*Source of water*		*Total*
	In-house connection	*Public source*	
Below 25	-	1[.6]	1[.2]
25-50	8[3.0]	7[4.0]	15[3.4]
50-75	107[40.5]	101[57.8]	208[47.4]
75-100	56[21.3]	34[19.4]	90[20.5]
100-125	63[23.8]	20[11.4]	83[18.9]
125-150	16[6.0]	9[5.1]	25[5.7]
150 and above	14[5.4]	3[1.7]	17[3.9]
Total	264[100.0]	175[100.0]	439[100.0]

Source: Survey Data, 2005.

Note : Figure in the brackets indicates column percentage.

water connection consume more water when compared to the households depending on water from common sources.

Null Hypothesis (H_0):

There is no significant difference in the metro-water consumption of households with in-house water connection and households without in-house water connection.

Alternative Hypothesis (H_a):

There is a significant difference in the metro-water consumption of households with in-house water connection and households without in-house water connection (Table 6.18).

Table 6.18. Chi-Square Test Results

Tests	*Value*	*df*	*Asymp. Sig. (2-sided)*
Pearson Chi-Square	77.111(a)	6	.000
Likelihood Ratio	89.966	6	.000
Linear-by-Linear Association	11.393	1	.001
N of Valid Cases	439		

a 0 cells (.0%) have expected count less than 5. The minimum expected count is 12.22.

The tabulated value χ^2 at 1 per cent level of probability is 22.5 for 6 degrees of freedom. However, the calculated value is higher than the tabulated value. The value is higher than the tabulated value. The value is significant. Hence it can be concluded that there exist a significant difference between the metro-water consumption of households with in-house water connection and households without in-house water connection. Therefore the null hypothesis is rejected and the alternative hypothesis is accepted.

The literacy status of the heads of the households classified according to the per capita per day consumption of water is shown in Table 6.19.

The general outcome is that at the lower water consumption levels, the illiterates share a greater proportion while the literates have more percentage share in the higher water consumption levels except for a minor difference at the highest level. Hence it is obvious that the literates consume more water when compared to the illiterates.

Table 6. 19. Classification of Literacy Status of the Respondent according to the Total Per Capita Per Day Consumption of Water

LCPD	*Literacy status*		*Total*
	Literate	*Illiterate*	
Below 25	-	1[1.1]	1[.2]
25-50	16[4.5]	3[3.2]	19[4.2]
50-75	157[44.2]	55[57.9]	212[47.1]
75-100	74[20.8]	18[18.9]	92[20.4]
100-125	73[20.6]	11[11.6]	84[18.7]
125-150	22[6.2]	3[3.2]	25[5.6]
150 and above	13[3.7]	4[4.2]	17[3.8]
Total	355[100.0]	95[100.0]	450[100.0]

Source: Survey Data, 2005.
Note : Figure in the brackets indicates column percentage.

The per capita per day water consumption of the households according to the area of the house is shown in Table 6.20.

There is no distinct relationship between the floor area of the house and daily water consumption. The basic feature is that at the lower water consumption levels, households with houses of lesser floor areas occupy more percentage share when compared to the households with houses of larger floor. The reverse is true in the case of higher water consumption levels. Hence it is clear that larger the floor area more will be the level of water consumption. Water consumption also depends upon the various auxiliary attachments in the houses and the availability of water-using equipments. It is found that

Table 6. 20. Classification of Area of Houses according to Total per Capita per Day Consumption of Water

LCPD	*Total Area of House (in sq.ft.)*						*Total*
	Below 300	*300-600*	*600-900*	*900-1200*	*1200-1500*	*1500 and above*	
Below 25	-	1[.7]	-	-	-	-	1[.2]
25-50	1[4.8]	4[2.8]	7[5.6]	4[5.5]	1[1.7]	2[6.5]	19 [4.2]
50-75	11[52.4]	83[58.5]	56[45.2]	26[35.6]	22[37.3]	14[45.2]	212[47.1]
75-100	5[23.8]	29[20.4]	18[14.5]	19[26.0]	14[23.7]	7[22.6]	92[20.4]
100-125	3[14.3]	12[8.5]	34[27.4]	13[17.8]	16[27.1]	6[19.4]	84[18.7]
125-150	1[4.8]	117.7]	3[2.4]	7[9.6]	2[3.4]	1[3.2]	25[5.6]
150 and above		2[1.4]	6[4.8]	4[5.5]	4[6.8]	1[3.2]	17[3.8]
Total	21[100.0]	142[100.0]	124[100.0]	73[100.0]	59[100.0]	31[100.0]	450[100.0]

Source: Survey Data, 2005.

Note : Figure in the brackets indicates column percentage.

households using more showers, toilets, kitchen sinks, bathrooms and wash basins consume more quantity of water per capita per day, as compared to the households with less number of such facilities and also those households which are marked by the absence of such facilities.

The main difference in the consumption pattern of metro water and the total water by the sample households is that the poor rely heavily on public water for most of their consumption needs, while the rich households supplement their demand for water from private sources or by making investment in alternative strategies.

Distance of Residence and Water Consumption

Table 6.21 depicts a comparison of water consumption categories according to the distance between the residence and the street water sources.

There exists an inverse relationship between the distance of the residence from the street water sources and the per capita per day consumption of water. Moreover those who travel a long distance for fetching water are the people of the slum areas. These people are poor and they depend heavily on the street water supply for almost all uses. Even then the trend is that the households consuming more water are traveling shorter distances to fetch water.

In order to get a clear picture of water consumption level the per capita per day consumption of water is compared with the time spent for fetching water. The results are furnished in Table 6.22.

It is evident that the travelling time taken for water collection does not make any significant impact on water consumption. Many of the households with more per capita per day consumption of water are found in the category of taking more travelling time to collect water.

Table 6.21. Total Per Capita per Day Consumption of Water and Distance between Residence and Street Water Source

LCPD	*Distance between residence and street water source*					*Total*
	Less than 50	*50-100*	*100-150*	*150-200*	*200 and above*	
Below 25	-	-	-	1 [4.5]	-	1 [0.6]
25-50	-	3 [3.3]	1 [2.8]	1 [4.5]	233.3]	7 [4.0]
50-75	14 [73.7]	55 [59.8]	20 [55.6]	8 [36.4]	4 [66.7]	101[57.7]
75-100	2 [10.5]	16 [17.4]	11[30.6]	5 [22.7]	-	34 [19.4]
100-125	2 [10.5]	13[14.1]	1 [2.8]	5 [22.7]	-	21 [12.0]
125-150	1 [5.3]	5 [5.4]	2 [5.6]	1 [4.5]	-	9[5.1]
150 and above	-	-	1[2.8]	1 [4.5]	-	2 [1.1]
Total	19 [100.0]	92[100.0]	36[100.0]	22[100.0]	6[100.0]	175[100.0]

Source: Survey Data, 2005.
Note : Figure in the brackets indicates column percentage.

Table 6. 22. Total Per Capita Per Day Consumption of Water and Travelling Time for Water Collection

LCPD	Travelling time for water collection								Total
	Less than 0.30	0.30-1.00	1-1.30	1.30-2.00	2.00-2.30	2.00-3.00	3.00-3.30	3.30 and above	
Below 25	-	-	-	-	-	-	-	1 [5.9]	1 [0.6]
25-50	1 [50.0]	1 [3.8]	1 [2.6]	2 [4.0]	1 [3.7]			1 [5.9]	7 [4.0]
50-75	1 [50.0]	17 [65.4]	26 [68.4]	28 [56.0]	15 [55.6]	2 [20.0]	4 [80.0]	8 [47.1]	101 [57.7]
75-100	-	4 [15.4]	7 [18.4]	10 [20.0]	5 [18.5]	3 [30.0]	1 [20.0]	4 [23.5]	34 [19.4]
100-125	-	2 [7.7]	4 [10.5]	6 [12.0]	4 [14.8]	3 [30.0]	-	2 [11.8]	21 [12.0]
125-150	-	2 [7.7]	-	3 [6.0]	2 [7.4]	1 [10.0]	-	1 [5.9]	9 [5.1]
150 and above	-	-	-	1 [2.0]	-	1 [10.0]	-	-	2 [1.1]
Total	2 [100.0]	26 [100.0]	38 [100.0]	50 [100.0]	27 [100.0]	10 [100.0]	5 [100.0]	17 [100.0]	175 [100.0]

Source: Survey Data, 2005.
Note : Figure in the brackets indicates column percentage.

Table 6. 23. Total Per Capita per Day Consumption of Water and Waiting Time for Collecting Water

LCPD	*Waiting time per head at present*					*Total*
	.30	*1.00*	*1.30*	*2.00*	*2.30*	
Below 25	-	-	1 [4.5]	-	-	1[.6]
25-50	7 [5.8]	-	-	-	-	7 [4.0]
50-75	72 [60.0]	14 [48.3]	13 [59.1]	2 [66.7]	-	101[57.7]
75-100	24 [20.0]	7 [24.1]	3 [13.6]	-	-	34[19.4]
100-125	9 [7.5]	8 [27.6]	2 [9.1]	1 [33.3]	1 [100.0]	21 [12.0]
125-150	6 [5.0]	-	3 [13.6]	-	-	9[5.1]
150 and above	2 [1.7]	-	-	-	-	2 [1.1]
Total	120[100.0]	29[100.0]	22[100.0]	3 [100.0]	1 [100.0]	175[100.0]

Source: Survey Data, 2005.
Note : Figure in the brackets indicates column percentage.

Table 6.23 shows the classification of per capita per day consumption of water by households, according to the waiting time for collecting water.

Waiting time also does not make any significant impact on the extent of water consumption. It is because of this reason that a large majority of the households consuming more water are found in the category of spending more waiting time.

Determinants of Household Water Consumption in Chennai City

In order to understand the determinants of water consumption by different categories of households, the per capita metro water consumption at the household level has been taken as the dependent variable. Hence the analysis is based on the factors influencing the per capita water consumption of the households in Chennai City. It has been decided to run a multiple regression model for metro water consumption. Most of the selected independent variables are separately compared with dependent variable.

The model relates to the number of households that consume water from metro water supply sources. The data described is a cross section data of 450 households in the selected ten Zones across Chennai City.

Null Hypothesis (H_o):

Household income has no influence on the volume of metro water consumption.

Alternative Hypothesis (H_a):

Household income has influence on the volume of metro water consumption.

MODEL

In this model, nine variables covering various aspects of water consumption are regressed against the dependent variable of Per Capita Metro Water Consumption per Day (Mlpcd). The variables incorporated in the model are as follows:

MLPCD → Per capita metro water consumption per day.
HSIZE → Size of the Household
MHDINCOME → Monthly Household income
NSHRS → Number of Showers
NFLTOLTS → Number of Flushing Toilets
NKITSINK → Number of Kitchen Sinks
DRESWTRSCRCE → Distance between Residence and Water Supply Point
DWATPAY → Details of Water Charge Payment in the Previous Year
AGEWATFETTIME → Average Time spent for Fetching Water
WATERSOURCE → Source of Water Supply

The multiple linear regression models have the following general formulation.

$$y_t = \beta_1 + \beta_2 Xt_2 + \dots\dots\dots\dots\dots\dots\dots\dots \beta_k t_k + U_t$$

The subscript refers to the observation number and it varies from 1 to n. The dependent variable y_t is the per capita metro water consumption per day. There are k independent variables and hence 'k' is unknown regression coefficient. Based on the econometric expression of the regression equation, the functional form of the multiple linear models for the determination of the per capita metro water consumption per day is evolved.

$$MLPCD = \beta_1 + \beta_2 HSIZE + \beta_2 MHDINCOME + \beta_3 NSHRS + \beta_4 NFLTOLTS + \beta_5 NKITSINK + \beta_6 DRESWTRSRCE + \beta_7 DWATPAY + \beta_8 AGEWATFETTIME + \beta_9 WATERSUPPLYSOURCE + U$$

Table 6.24. Descriptive Statistics

S.No	*Variables used*	*Mean*	*Std. Deviation*	*N*
1	Per capita average metro water consumption per day (in liters) β1	71.86	36.31	439
2	Size of the Household (Nos.) β2	4.32	1.45	439
3	Monthly Household Income (in Rs.) β3	8369.25	7668.78	439
4	Number of Showers β4	.27	.47	439
5	Number of Flushing Toilets β5	1.30	.63	439
6	Number of Kitchen Sinks β6	.65	.77	439
7	Distance between Residence and Water Supply Point (in meters) β7	48.16	68.78	439
8	Water Charge Payment in the Previous Year (in Rs.) β8	356.62	307.38	439
9	Average Time Spent for Fetching Water β9	1.59	2.42	439
10	Source of Water Supply β10	1.36	.53	439

Source: Survey Data.

Table 6.25. Inter Pearson Correlation Matrix

Variables used		*Per capita average metro water consumption per day*	*Size of the household*	*Monthly household income*	*Number of showers*	*Number of flushing toilets*
Per capita average metro water consumption per day (in litres)	β 1	1.000				
Size of the Household (Nos.)	β 2	.238	1.000			
Monthly Household Income (in Rs.)	β 3	-.186	.049	1.000		
Number of Showers	β 4	-.258	.007	.468	1.000	
Number of Flushing Toilets	β 5	-.227	.082	.459	.344	1.000
Number of Kitchen Sinks	β 6	-.271	-.024	.542	.527	.632
Distance between Residence and Water Supply Point (in meters)	β 7	.140	.068	-.376	-.313	-.373
Water Charge Payment in the Previous Year (in Rs.)	β 8	-.086	-.039	.415	.346	.431
Average Time Spent for Fetching Water	β 9	.230	.228	-.347	-.275	-.329
Source of Water Supply	β 10	.248	.055	-.391	-.311	-.390

(Contd.)

(Contd. Table 6.25)

Variables used		Number of kitchen sink	Distance between residence and water source	Pay for water bill in the last year	Average time spent for fetching water	Source of water
Per capita average metro water consumption per day (in litres)	β 1					
Size of the Household (Nos.)	β 2					
Monthly Household Income (in Rs.)	β 3					
Number of Showers	β 4					
Number of Flushing Toilets	β 5					
Number of Kitchen Sinks	β 6	1.000				
Distance between Residence and Water Supply Point (in meters)	β 7	-.446	1.000			
Water Charge Payment in the Previous Year (in Rs.)	β 8	.524	-.814	1.000		
Average Time Spent for Fetching Water	β 9	-.409	.876	-.761	1.000	
Source of Water Supply	β 10	-.477	.833	-.791	.769	1.000

Source: Survey Data, 2005.

The descriptive statistics of the variables used in the model are given in Table 6.24.

The theoretical relationships between the dependent variable and the independent variables have already been discussed. In order to understand the interrelationship between the predictor variables chosen, and the criterion variable an inter-correlation matrix has been constructed and it is given in Table 6.25.

The independent variables like monthly household incomes, number of showers, number of flushing toilets, number of kitchen sinks, households paying water bill all are negatively related to the per capita metro water consumption per day of the sample households. Hence it is quite clear that the attributes of better-off households are negatively related to the dependent variable.

The independent variables like household size, distance between residence and metro water supply source, average time spent for water collection and source of water are positively correlated to the per capita metro water consumption per day. The positively correlated variables are the attributes of poorer sections of the sample households. The inter-correlation between the independent variables is not very high, and therefore the problem of multi-collinearity does not arise.

Based on the significance of correlation, six predicator variables (independent variables) have emerged in the category of 'entered' or 'removed'. They include the number of kitchen sinks, household size, number of showers, source of water, payment of water bill and the number of flushing toilets. These independent variables are regressed against the criterion variable of per capita metro water consumption per day and the coefficients estimated with the help of Ordinary Least Squares (OLS) step-wise model. The summary results of the model are furnished in Table 6.26.

Table 6.26. Model Summary

Model	*R*	*R Square*	*Adjusted R Square*	*Std. Error of the Estimate*	*Change Statistics*					*Durbin-Watson*
					R Square Change	*F Change*	*Df1*	*Df2*	*Sig. F Change*	
1	.271	.074	.072	34.98758	.074	35.617	1	448	.000	
2	.357	.127	.124	33.99402	.054	27.571	1	447	.000	
3	.384	.147	.141	33.64441	.020	10.338	1	446	.001	
4	.400	.160	.152	33.42820	.013	6.788	1	445	.009	
5	.468	.219	.210	32.26697	.059	33.606	1	444	.000	
6	.480	.230	.220	32.06829	.011	6.519	1	443	.011	1.759

The value of adjusted R^2 is 0.220, indicating that 22 per cent of the variance of the per capita metro water consumption per day is explained collectively by the six variables entered and removed in the model.

a Predictors: (Constant), Number of kitchen sink
b Predictors: (Constant), Number of kitchen sink, Size of the household
c Predictors: (Constant), Number of kitchen sink, Size of the household, Number of showers
d Predictors: (Constant), Number of kitchen sink, Size of the household, Number of showers, Source of water
e Predictors: (Constant), Number of kitchen sink, Size of the household, Number of showers, Source of water, Pay for water bill in the last year
f Predictors: (Constant), Number of kitchen sink, Size of the household, Number of showers, Source of water, Pay for water bill in the last year, Number of flushing toilets
g Dependent Variable: Per capita average metro water consumption per day

The results indicate that a 1 per cent increase in per capita metro water consumption per day is associated with 7 per cent decrease in kitchen sinks, 5 per cent increase in the household size, 2 per cent decrease in the number of showers and 1 per cent decrease in the number of flushing toilets. The variables rejected in the model include monthly household income, distance between residence and water source and the average time taken for fetching water. The Durbin Watson statistic is estimated as 1.759(d). Hence the selected variables explain the determination of per capita metro water consumption per day by the household. Since household income is rejected in the step-wise OLS multiple regression model, it confirms that household income has no influence on the volume of metro water consumption. Hence it is concluded that the null hypothesis is accepted and the alternative hypothesis is rejected.

Table 6.27. Values of coefficients, standard error, â coefficients, t value and the significance

Model		*Unstandardized Coefficients*		*Standardized Coefficients*		
		B	*Std. Error*	*Beta*	*T*	*Sig.*
1	(Constant)	80.258	2.168		37.020	.000
	Number of kitchen sink	-12.853	2.154	-.271	-5.968	.000
2	(Constant)	54.923	5.265		10.432	.000
	Number of kitchen sink	-12.594	2.093	-.266	-6.017	.000
	Size of the household	5.823	1.109	.232	5.251	.000
3	(Constant)	55.331	5.212		10.615	.000
	Number of kitchen sink	-8.461	2.438	-.179	-3.470	.001
	Size of the household	5.902	1.098	.235	5.376	.000
	Number of showers	-12.830	3.990	-.165	-3.215	.001
4	(Constant)	41.936	7.297		5.747	.000
	Number of kitchen sink	-5.816	2.627	-.123	-2.214	.027
	Size of the household	5.755	1.092	.229	5.270	.000
	Number of showers	-11.993	3.978	-.155	-3.015	.003
	Source of water	8.878	3.408	.129	2.605	.009

(Contd.)

(Contd. Table 6.27)

Model		*Unstandardized Coefficients*		*Standardized Coefficients*	*T*	*Sig.*
		B	*Std. Error*	*Beta*		
5	(Constant)	.208	10.071		.021	.984
	Number of kitchen sink	-9.160	2.600	-.193	-3.523	.000
	Size of the household	5.722	1.054	.228	5.427	.000
	Number of showers	-13.246	3.846	-.171	-3.445	.001
	Source of water	28.701	4.745	.418	6.049	.000
	Water bill	4.889E-02	.008	.414	5.797	.000
6	(Constant)	6.705	10.328		.649	.517
	Number of kitchen sink	-5.459	2.963	-.115	-1.842	.066
	Size of the household	6.078	1.057	.242	5.750	.000
	Number of showers	-13.266	3.822	-.171	-3.471	.001
	Source of water	28.292	4.718	.412	5.996	.000
	Pay for water bill in the last year	5.072E-02	.008	.429	6.029	.000
	Number of flushing toilets	-8.140	3.188	-.140	-2.553	.011

A, Dependent Variable: Per capita average metro water consumption per day

Since 4 – 1.759 e" du, it can be concluded that (d = 2.241). There is no negative autocorrelation. Since the computed 'F' values of all the six predictor variables are higher than the values given in the F-distribution table at 1 per cent level of significance, the null hypothesis is rejected and the accepted variables used in the model are known to assert and influence the per capita metro water consumption per day of the selected households in Chennai City. The different values of coefficients, standard error, â coefficients, t value and the significance levels are furnished in Table 6.27.

The regression coefficient of the independent variables is many times its standard error; hence the estimates of regression coefficients have to be adjudged as significant. The beta coefficient indicates the relative contributions made by independent variables to estimating the dependent variable. All these tests associated with multiple regression models reveal that the selected independent variables influence the consumption demand of metro water in Chennai City.

7

Water Supply Unreliability and Alternative Strategies Adopted by Households in Chennai City

The previous chapter elaborated the nature and the extent of public water demand exhibited by the sample households in Chennai City. It is revealed that a majority of the households have inadequate access to safe Metro water. The middle and lower income households depend more on Metro water, while the rich, in order to avoid the unreliability of Metro water depend to a considerable extent upon private water sources to meet their water demand. Therefore, the present chapter intends to identify the indicators of the unreliability, the coping strategies across different sections of the households in their effort to face the problem of water unreliability and the extent of their demand for making the CMWSSB for maintaining and improving the metro water supply in Chennai City. The analysis of the present chapter is arranged in three sections. Water supply unreliability of the sampled households in Chennai City is discussed in Section

A, the compensatory strategies used by households in Chennai City in Section B and section C analyses the household's willingness to pay for existing and improved water supply in Chennai City.

SECTION – A

Water Supply Unreliability

An overview of the water supply unreliability is centred on the inadequate access of water in comparison to water needs of the households in Chennai City. The demand for water has been increasing much faster than its supply due to urbanization and industrialization. Population growth and improper supply side and demand side water management accelerate the deterioration of water resources. Munasinghe (1979) introduced the concept of unreliability in this respect. He used an *'unreliability measure'* in the case of power supply.

The technical know-how during the inception of power supply projects in developing countries was old. They were costly choices in terms of investment, similar to those in developed countries (Munasinghe, 1979). The prime objectives of infrastructures projects are to maximize benefits to society. Munashinge is critical of the traditional approach that has been based on the principle of minimizing supply cost required for a given level of reliability.

It means that user's demand should be taken into consideration during the inception and design processes of an infrastructure project or it is necessary to determine the level of service, which users are willing to accept. Reliability means that a facility will not fail during its service life (Humplick, 1991). From the user's point of view the facility should be accessible physically and affordable economically to satisfy their requisites.

In order to ensure reliable supply, the public services should be available without interruption. The quality of the public good, its intermittent supply, irregularity and unpredictability largely decide the extent of unreliability. According to Munashinge, most of the public services in developing countries are unreliable in nature. If the level of security of a public service is an endogenous variable, then the total cost of the infrastructure consists of infrastructure supply cost and cost of failure. Therefore, reliability is defined as a point where the total benefits are equal to the total costs.

Measurement of unreliability is a difficult process. It depends heavily on the cost of supplying the public good. Hence it is necessary to integrate data on demand in an iterative process. Moreover, the perception of unreliability changes from one economic player to another from one user to another, leading to overvaluation or undervaluation of the service. At the global level, the degree of unreliability is a function of the capability of an infrastructure system in handling the demands imposed on it, which depends on the type of activity requiring infrastructure as an input (Humplick, 1995).

The analysis of the unreliability of water supply is not much different from that of any other infrastructure services. Several factors influence water supply unreliability. They include limited water resources, inefficient network, lack of modernization, etc. The level of reliability of a service varies on a scale from 0 to 1, with the following values. 0<1=unreliability, and reliability=1 (Zerah, 2000).

In the cities of developing countries, lack of reliability is a characteristic of any water supply system which manifests itself in doubtful quality, intermittent supply, seasonal disruptions or all these. In normal conditions, the supply is unreliable. The estimation of total cost of unreliability is the cost of reducing water supply unreliability plus the cost of unreliability. The *'avoidance cost'* of unreliability is the investment in treatment plants, reservoirs, water lines, etc. The cost of unreliability, on the other hand, is the total cost borne by the user and the society. This represents productivity loss inclusive of the added costs related to water quality. There are thus two categories of costs, viz., cost of the compensatory strategies and the cost of the consequences of unreliability.

Characteristics of Unreliable Water Supply

Three important features characterize water supply unreliability. They are intermittent water supply, insufficient and irregular pressure and unpredictable service. Access to water does not mean a 24-hour water supply. In most developing countries, cities get water intermittently for a few hours everyday. It is not possible to predict whether water comes during the day or night.

When the households receive water from the piped connection, it took 7 to 20 minutes to fill 15 litre *Kodam* (pot). Sometimes it took more than 20 minutes to fill the *kodam*. This is characterized as insufficient and irregular supply of water. Due to technical problems, the duration of water service cannot be predicted. This is manifested in unpredictable water supply. Interruptions in the service are fairly frequent in summer months of May and June.

Studies on Water Supply Unreliability

Lee and Anas conducted the first study of the impact of unreliability on the cost of infrastructure use in the manufacturing sector in Nigeria. The study revealed the importance of undertaking investment in infrastructure and on the consequent inequalities resulting between small and large companies. The average price of water is lower for large companies due to the economies of scale. The results pointed out that industries with less than 100 employees paid more for water, while the opposite is true for large companies. Hence, the study recommends that the large companies have its own water supply and sell the surplus water to small companies.

The World Bank conducted four independent country specific studies. These studies concluded that households having in-house water connections made long-term investments to solve the problems of water quality and water supply unreliability. Those who depend on outside sources acquired water storage devices and hand pumps to cope with insufficient pressure in the water pipes. Households with indoor water connections also used water tanks, motor and water filters.

M.H. Zerach for Delhi conducted a more precise study on unreliability. The study inferred that water consumption of private colonies increases with the level of household equipment. The effect of the size of the family in water consumption was slightly significant. As the city grows, the challenge before the public authorities is to frame policies reflecting the importance of the social and economic costs of inadequate and insufficient water supply.

Private (domestic or foreign) investment in infrastructure is based on the likely rate of returns from the project over a period of time. The rate of returns is largely decided by the economically rational tariff structure. Thus, with a generous tariff and counter guarantees in the water supply sector measure of success in attracting investments. Generally, public investment in the water supply sector is grossly inadequate in the cities of developing countries. Zerah concludes in her study that lack of resources, poor management, tariff and subsidy constraints, financial difficulties, myopic strategies and regulatory framework were the prime causes of unreliability of water supply.

In-House Piped Water Connection

Service Coverage

It has already been noted in the previous chapter that the households' access to safe water is not uniform in their endowments. Nearly three fifths (60 per cent) of the households are in the category of in-house piped water connection. However, the percentage of households with metered water use is highly insignificant. Of the 264 households with in-house water connection, 59.1 per cent households are in South Chennai, as compared to 40.9 per cent of the same in North Chennai. The capital investment made for in-house water connection is different among different households. Some of the households are connected with in-house hand pipes and some others collected water in the underground sumps and then lifted it to the overhead tanks by means of electric pump sets. Those who do not have in-house water connection depend on the common public water delivery points. The years in which the households connected in-house water supply are shown in Table 7.1.

It is evident that nearly half (47.7 per cent) of the sample households connected in-house water supply between 1990 and 2000 and more than one-quarter (26.5 per cent) got the in-house water connection between 1980 and 1990.

The general trend visible is that the percentage of households with in-house water connection has been increasing decade after decade. Though more households are connected to in-house water supply, the problems of unreliability also increased simultaneously as a result of shortage of water supply. The problems of unreliability can be analyzed by examining the

Table 7.1. Years of In-House Water Connection

Years	*Households*	*Percentage*
Before 1970	4	1.5
1970-1980	22	8.3
1980-1990	70	26.5
1990-2000	126	47.7
2000 and above	42	15.9
Total	264	100.0

Source: Survey Data, 2005.

duration of water supply, lack of pressure in the pipelines and irregular supply of metro water.

Duration of Water Supply

The duration of Metro water supply vary differently in different divisions of Chennai City. There have been localized problems; hence, the supply of water is not uniform in different places. Supplying of water only for a short duration is one of the important indicators of unreliability. Despite huge variations in the availability of water per day in summer and winter months, the daily availability of water has been limited to three or four hours to the households with in-house water connection. This is shown in Table 7.2.

The responses of the respondents about the duration of Metro water supply in the previous date of the survey revealed that only 1.5 per cent of the households with in-house water connection received water for 12 hours or more. Such households are located very near the main water distribution

Table 7. 2. Duration of In-house Water Supply Per Day

Duration of water supply (in hours)	*Location*		*Total*
	North Chennai	*South Chennai*	
Less than 2	51 [47.2]	25 [16.0]	76 [28.8]
2 - 4	48 [44.4]	74 [47.4]	122 [46.2]
4 - 6	7 [6.5]	39 [25.0]	46 [17.4]
6 - 12	2 [1.9]	14 [9.0]	16 [6.1]
12 - 24		4 [2.6]	4 [1.5]
Total	108 [100.0]	156 [100.0]	264 [100.0]

Source: Survey Data, 2005.

Note : Figure in the brackets indicates column percentage.

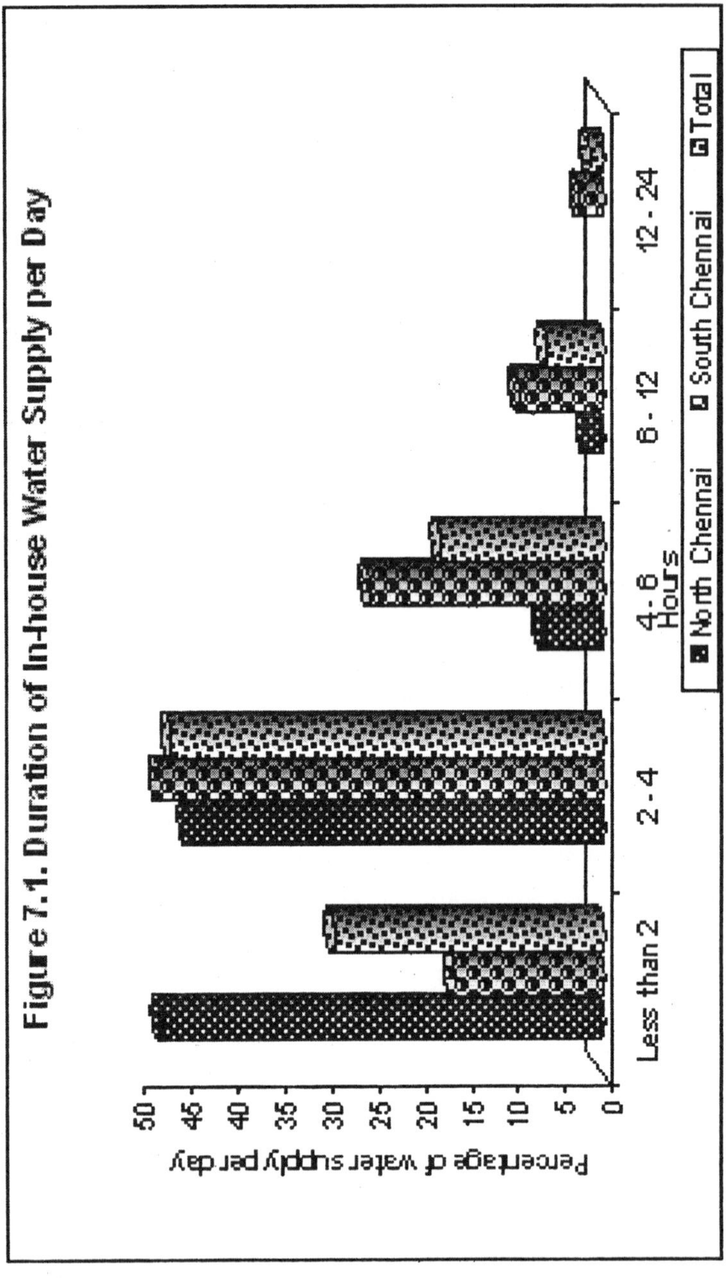

Figure 7.1. Duration of In-house Water Supply per Day

points. More than a quarter (28.8 per cent) of the households with in-house water connections obtained less than 2 hour's supply. Nearly half of the households (46.2 per cent) got 2 to 4 hours of water per day. On an average the duration of water supplied every day to the households connected to the in-house supply has been very low. Therefore, the average duration of daily supply in summer and winter months does not make any big difference.

The proportion of households with the highest duration of 12 hours and more per day is unequally distributed. All such households are located in South Chennai. On the other hand, nearly half of the households with in-house water connection in North Chennai represent in the category of the lowest duration of Metro water supply of less than 2 hours per day. From this, it is very clear that not only percentage of households with in-house water connection is less in North Chennai but that also they receive lower average duration of Metro water supply compared to the same type of households in South Chennai.

Generally the residences of the affluent people it is averred to get more hours of water supply because of their ability to influence the authorities. In order to understand the relationship between affluence and availability of better service, monthly income of the sample households is compared with the duration of water supply per day. This is furnished in Table 7.3. The comparison of household incomes with the duration of water supply to those households with in-house water connection reveals that the higher income households get more hours of water supply. Nevertheless, it is found that 46.2 per cent of the households with in-house water connection have only 2 to 4 hours of water supply everyday. Among them the higher income groups of Rs.12500 and above per month constitute 35.2 per cent. The Pearson's correlation between monthly household income and the duration of water supply per day of the households with in-house water connection is computed as 0.155,

Table 7. 3. Household Monthly Income and Duration of Daily In-house Water Supply (in Hours)

Monthly Income (Rs.)	*Duration of Water Supply (in Hours)*					*Total*
	0 to 2	*2 to 4*	*4 to 6*	*6 to 12*	*12 to 24*	
Below 2500	6 [7.9]	10 [8.2]	4 [8.7]			20 [7.6]
2500 - 5000	22 [28.9]	25 [20.5]	4 [8.7]	3 [18.8]		54 [20.5]
5000 - 7500	15 [19.7]	16 [13.1]	4 [8.7]	2 [12.5]	1 [25.0]	38 [14.4]
7500 - 10000	7 [9.2]	19 [15.6]	10 [21.7]	2 [12.5]	1 [25.0]	39 [14.8]
10000 - 12500		9 [7.4]	1 [2.2]			10 [3.8]
12500 - 15000	9 [11.8]	11 [9.0]	9 [19.6]	1 [6.3]		30 [11.4]
15000 and above	17 [22.4]	32 [26.2]	14 [30.4]	8 [50.0]	2 [50.0]	73 [27.7]
Total	76 [100.0]	122 [100.0]	46 [100.0]	16 [100.0]	4 [100.0]	264 [100.0]

Source: Survey Data, 2005.
Note : Figure in the brackets indicates column percentage.

which shows a relatively low degree of positive correlation between these two variables.

Hence the assumption that more affluent areas benefited better service of water supply is not substantiated. This agrees with the connotation that income would play a decisive role in the choice of compensatory strategies rather than working for the betterment of the common property resource.

Low Water Pressure

Low water pressure is one of the important determinants of unreliability. The estimation of water pressure is a difficult job than measuring the number of hours of distribution. The availability of water in the short duration of supply to the households with in-house connection is largely based on water pressure. Normally, water pressure at taps is estimated on the basis of distribution characteristics.

Table 7.4 shows the classification of water pressure at taps of households with in-house water connection.

Table 7.4. Nature of the Pressure Variation in In-House Connections

Nature of Pressure Variation in the in-house Water Connection	*Location*		*Total*
	North Chennai	*South Chennai*	
Good Pressure (Less than 2 minutes)	1[.9]	41[26.3]	42[15.9]
Average Pressure (between 3 and 6 minutes)	45[41.7]	80[51.3]	125[47.3]
Poor Pressure (between 7 and 20 minutes)	61[56.5]	33[21.2]	94[35.6]
No Pressure (more than 20 minutes)	1[.9]	2[1.3]	3[1.1]
Total	108[100.0]	156[100.0]	264[100.0]

Source: Survey Data, 2005.

Note : Figure in the brackets indicates column percentage.

It is revealed that the proportion of households with good pressure is a little over 15 per cent. A majority of the households (47.3 per cent) received water at an average pressure, followed by those with poor pressure, accounting for 35.6 per cent of the households with in-house water connection.

The inequalities due to pressure are more severe in North Chennai because good pressure accounts for only 0.9 per cent of the households. The percentage of households with poor pressure is 56.5 and those with average pressure exceed two-fifths (41.7 per cent) of the households with in-house connections.

The duration of water availability for the households with in-house water connection is not only short but also attributed to low and irregular water pressure. Water pressure in South Chennai is better placed when compared to North Chennai, because the main water distribution stations are located in South Chennai and the pipe network system of North Chennai is obsolete without proper replacement and repair. Thus, short duration of water availability without adequate pressure reveals that the households receive in-house water connection with high degree of unreliability, and therefore, they seek alternative strategies to cope with acute shortage of water. Interruption of water services is very serious in summer. Though the Government tried to meet the situation with increased mobile water-tanker lorries, the hardships of households would be tough. Mobile tanker lorry services are meant for the areas of complete absence of piped water supply. Seldom do the rich households try to face the hardship of unreliability, since alternative choices of private water supplies are available to them.

Time of Water Supply

The most striking dimension of unreliable water supply is its unpredictability. The time of water availability is not clearly known to the households. They behave on the basis of their past experiences. If water was available at night in the previous day they expect the same will repeat next day too. The time of water availability of households with in-house connection in the previous day of the survey is shown in Table 7.5.

It is clear that more than half (54.2 per cent) of the households get Metro water in the morning. This is almost the same for both South Chennai and North Chennai. Almost one-fourth of the households (24.2 per cent) with in-house water connection get water at night. Both morning and night are the common time of supply, because such time is convenient for the members in the households to make the necessary

Table 7.5. Time of In-house Water Supply

Water supply	*Location*		*Total*
	North Chennai	*South Chennai*	
Morning	57 (39.9) [52.8]	86 (60.1) [55.1]	143 (100.0) 54.2]
After noon	3 (21.4) [2.8]	11 (78.6) [7.1]	14 (100.0) [5.3]
Evening	6 (25.0) [5.6]	18 (75.0) [11.5]	24 (100.0) [9.1]
Night	38 (59.4) [35.2]	26 (40.6) [16.7]	64 (100.0) [24.2]
Any time	4 (21.1) [3.7]	15 (78.9) [9.6]	19 (100.0) [7.2]
Total	108 (40.9) [100.0]	156 (59.1) [100.0]	264 (100.0) [100.0]

Source: Survey Data, 2005.
Note : Figure in the parenthesis indicates row percentage.
Figure in the brackets indicates column percentage.

arrangements. But on the other hand, this will give enormous hardship to them in the form of inconvenience during the office-going time and sleeplessness at night.

The sample area of Chennai City consists of independent houses, multi-storied buildings, apartments and flats. Among the 264 households with in-house water connections, 26.8 per cent share the water with others. The water sharing details are given in Table 7.6.

Table 7.6. Sharing of in-house Water Supply

No. of Households Sharing in-house water	*Sample households*	*Percentage*
One	9	12.7
Two	18	25.4
Three	13	18.3
Four and more	31	43.7
Total	71	100.0

Source: Survey Data, 2005.

Out of the 71 households 43.7 per cent shared water along with four or more than four households. It is evident that the households having in-house water connection face severe unreliability problems like short duration of supply, low pressure and un-predictability. Therefore, they invest in equipments and infrastructural facilities that help to meet the contingencies arising out of water unreliability. When the degree of unreliability is rapid, they purchase water from private sources.

Public Water Source (Metro Water)

The Metro Water Board takes the responsibility of supplying water to those households that do not have in-house water connections. Such households take water from any one of the public water supply points. The Board supplies water through street stand-posts, street hand pumps, fixed water tanks and mobile water tanker lorries. In an attempt to gain some perspectives of the accessibility on public water to the households, the survey attempted to capture the distance between the nearest public water source and the residences of the user households. Understandably, such information would be indicative of the quality of life and provide a better understanding of the nature of hardships faced by the households without in-house water connections. The unreliability concept is more relevant to the households without in-house water connections. The proportion of such households, which depend entirely on street common water sources, is 38.67 per cent or 175 of the 450 households. The coverage of unreliability of the households without in-house water connections is much broader than the households with in-house water connections. Important indicators of unreliability of the households without in-house water connection include water pressure at public taps, water availability, the number of households sharing water from the distribution points, time of water availability, waiting time, distance between residence and distribution point, etc. The households without in-house water connections have to depend on the public water distribution points for their water needs. Unlike the households with in-house water connections, these households have to share a single distribution point along with many others.

Nature of Water Pressure in the Public Water Sources

Water pressure at the taps in the distribution points is one of the important determinants of unreliability. When water pressure at the tap is low, the depending households have to spend more time for water collection. Table 7.7 shows the nature of water pressure at the tap for the households without in-house water connection.

The measure of water pressure is the same for both households with in-house water connection and without in-house water connection. It is estimated on the basis of the time

Table 7.7. Water Pressure in the Common Distribution Points

Nature of pressure	*Location*		*Total*
	North Chennai	*South Chennai*	
Good pressure (less than 2 minutes)	14 [12.4]	7 [11.3]	21 [12.0]
Average pressure (between 3 and 6 minutes)	46 [40.7]	39 [62.9]	85 [48.6]
Poor pressure (between 7 and 20 minutes)	48 [42.5]	16 [25.8]	64 [36.6]
No pressure (more than 20 minutes)	5 [4.4]		5 [2.9]
Total	113 [100.0]	62 [100.0]	175 [100.0]

Source: Survey Data, 2005.
Note : Figure in the brackets indicates column percentage.

required to fill up one 15-litre *kodam* with water. As expected, North Chennai has 64.6 per cent of the total households without in-house water connections. The quality of water pressure in North Chennai is comparatively poor. The proportion of the households with good pressure is 12.4 per cent in North Chennai which is a little higher than that of South Chennai. On an average nearly half (48.6 per cent) of the households without in-house water connection have average water pressure. The water pressure indicator reveals that 42.5 per cent of the households in North Chennai face poor pressure as compared to 25.8 per cent of households in South Chennai. No household for South Chennai is represented in the category of '*no pressure*'. However, 4.4 per cent of the households in North Chennai face '*no pressure*' at their public water taps. It is found during the survey that the water pressure was not uniform throughout the duration of water availability every day. Only a few households in South Chennai enjoyed constant pressure during the time of water availability, because such water distribution points are located near water distribution stations.

In addition to the problem of low pressure, the households depend on water from street sources frequently face short-term breakdown in water supply system due to technical snags and repairs. They also face the problems of irregular pressure. The problems of repair lead the users to mobilize funds for meeting the maintenance cost. In some of the distribution points, the leader of the users collected some fee ranging from Rs.25 to Rs.50 per month in order to meet the contingencies. Another pathetic situation during the repair is that the households have to travel long distance to fetch water.

Moreover, each distribution point has a group of users. They will not allow anybody from outside when the duration of water supply is short. Even among the users, the distribution of water is based on priority. The priority is fixed either on their position in the queue or the number of pots *(Kodams)* allotted to each households, irrespective of the number of members in the households. It is learnt that the hardships of households could be severe during the summer months. The quality and quantity of water supplied and its management are considered as important factors, influencing electoral verdict in the city.

Availability of Street Water

The respondents were asked about the availability of street water in terms of the number of days in the previous week of the survey. It was reported that all the households are not getting water every day. The availability of public water to the households in a week is shown in Table 7. 8.

The proportion of households getting continuous water supply in all the days in a week is very meager. They account for only 2.3 per cent of the households without in-house water connection. It is revealed that nearly three fifths (57.7 per cent) of the households were supplied water for four days in a week.

Table.7.8. Availability of Public Water

Number of days in a week	*Location* North Chennai	South Chennai	*Total*
One		1 [1.6]	1 [.6]
Two	2 [1.8]	1 [1.6]	3 [1.7]
Three	32 [28.3]	28 [45.2]	60 [34.3]
Four	71 [62.8]	30 [48.4]	101 [57.7]
Five	4 [3.5]	1 [1.6]	5 [2.9]
Six	1 [.9]		1 [.6]
Seven	3 [2.7]	1 [1.6]	4 [2.3]
Total	113 [100.0]	62 [100.0]	175 [100.0]

Source: Survey Data, 2005.
Note : Figure in the brackets indicates column percentage

In North Chennai, the proportion of households, which are supplied water four days in a week, are to the extent of 62.8

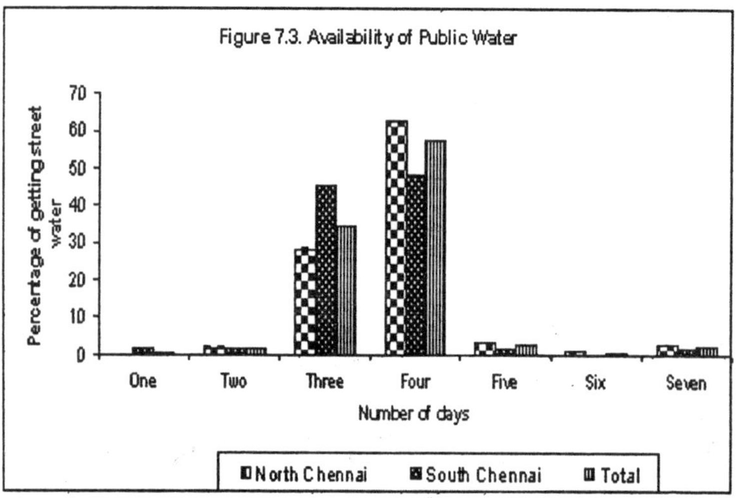

Table 7.9. Households Monthly Income and Availability of Public Water

Monthly Income (Rs.)	*Number of days in a week*							*Total*
	One	*Two*	*Three*	*Four*	*Five*	*Six*	*Seven*	
Below 2500		1 [33.3]	24 [40.0]	35 [34.7]	5 [100.0]			65 [37.1]
2500 - 5000		2 [66.7]	22 [36.7]	46 [45.5]		1 [100.0]	3 [75.0]	74 [42.3]
5000 - 7500			12 20.0]	13 [12.9]				25 [14.3]
7500 - 10000			2 [3.3]	3 [3.0]				5 [2.9]
10000 - 12500				1 [1.0]			1 [25.0]	2 [1.1]
12500 - 15000				1 [1.0]				1 [.6]
15000 and above	1 [100.0]			2 [2.0]				3 [1.7]
Total	1 [100.0]	3 [100.0]	60 [100.0]	101 [100.0]	5 [100.0]	1 [100.0]	4 [100.0]	175 [100.0]

Source: Survey Data, 2005.
Note : Figure in the brackets indicates column percentage.

per cent. Such households in South Chennai receive water for lesser number of days in a week, when compared to the similar households in North Chennai. Nevertheless, the variations are not considerable.

Monthly Income and Water Availability

The link between monthly household income and the public water availability to the households without in-house water connection is shown in Table 7.9.

It indicates that four-fifths (79.4 per cent) of the households without in-house water connection belong to the monthly income category of less than Rs.5000. About three-fifths (57.7 per cent) of the sample households without in-house water connection get water supply for four days in a week. A majority of them represent the lowest monthly income categories of below Rs.2500 and Rs.2500-5000. It indicates that the lower income groups face the problems of unreliable water supply severely when compared to the rich.

The intensity of unreliable water supply is severe for lower income households. Not only do they face acute shortages of water but also they face quantity restrictions. Further, it can be surmised that a very high opportunity cost is incurred by these households for the time spent on fetching water. The quantitative restrictions regarding water delivery are in the form of fixing the number of *Kodams* per household and the conditions of standing in the queue for long, on a first come first served priority. In this process of fetching water, the entire family members may be involved.

Time of Water Availability

Intermittent water supply is termed as one of the important indicators of unreliable water supply. The interruption in water supply is frequent for the households, which fetch water from the street sources. The strain of fetching water at the time of unpredictable water availability causes households to forgo their routine work, reduce water consumption and to search for alternative water sources. Table 7.10 gives the time of availability of water for the households taking water from outside their residence.

The Metro water generally supplies water either in the night or in the morning by taking into account the interest of the people going outside for work, in the daytime. Out of the 175 households fetching water from outside their residence, 36.6 per cent collect water in the morning and 29.1 per cent in the night. But 22.9 per cent of the households are not able to say with certain the time of water availability. The precarious time of water availability add fuel to the problem of water

Table 7.10. Time of Public Water Availability

Time	*Location*		*Total*
	North Chennai	*South Chennai*	
Morning	46 (71.9) [40.7]	18 (28.1) [29.0]	64 (100.0) [36.6]
After noon	4 (50.0) [3.5]	4 (50.0) [6.5]	8 (100.0) [4.6]
Evening	6 (50.0) [5.3]	6 (50.0) [9.7]	12 (100.0) [6.9]
Night	35 (68.6) [31.0]	16 (31.4) [25.8]	51 (100.0) [29.1]
Any time	22 (55.0) [19.5]	18 (45.0) [29.0]	40 (100.0) [22.9]
Total	113 (64.6) [100.0]	62 (35.4) [100.0]	175 (100.0) [100.0]

Source: Survey Data, 2005.

Note : Figure in the parenthesis indicates row percentage.
Figure in the brackets indicates column percentage.

unreliability. A comparison of the households in North Chennai and South Chennai revealed that over 70 per cent of the households in North Chennai reported water availability both in the morning and night. However, unpredictability of water availability is extreme in South Chennai because 29 per cent of the households getting water from common source do not know the time of water availability.

Time Spent for Collecting Water

Based on the distance between the residence and the nearest public water distribution point and the number of *kodam* of water fetched, the total distance travelled for fetching water by each household is estimated. The estimation of the time spent on water collection includes the waiting time in the queue. The mean time of travel for water collection by the households per day is estimated at 0.2503 hours, with a standard deviation of 0.11289. The mean waiting time per household per day is computed as 2.2691 hours with a standard deviation of .69088. Therefore, it becomes very clear that households without in-house water connection spent considerable time every day for collecting water.

It is generally perceived that people with better economic status would not spend much time in procuring water. Instead, they seek alternative sources of water. Hence the household income of those without in-house water connection is compared with the travel time required for collecting water. The details are furnished in Table.7.11.

It is evident that 45 of the 175 households (25.7 per cent) without in-house water connection spend less than 10 minutes per day for collecting water. Further, 41 of households (23.4 per cent) have spent 10-20 minutes each time for water collection, 45 households (25.7 per cent) have spent 20-30 minutes and 44 households (25.1 per cent) have spent 30 minutes and above.

Table 7.11. Distribution of Monthly Household Income according to Per Capita Travel Time Taken to Fetch Water

Households Monthly Income (Rs.)	*Travel time per head per household*				*Total*
	Less than 10	*10-20*	*20-30*	*30 and above*	
Below 2500	**16** (24.6) [35.6]	**16** (24.6) [39.0]	**22** (33.8) [48.9]	**11** (16.9) [25.0]	**65** (100.0) [37.1]
2500 - 5000	**22** (29.7) [48.9]	**15** (20.3) [36.6]	**15** (20.3) [33.3]	**22** (29.7) [50.0]	**74** (100.0) [42.3]
5000 - 7500	**4** (16.0) [8.9]	**7** (28.0) [17.1]	**5** (20.0) [11.1]	**9** (36.0) [20.5]	**25** (100.0) [14.3]
7500 - 10000	**1** (20.0) [2.2]	**2** (40.0) [4.9]	**1** (20.0) [2.2]	**1** (20.0) [2.3]	**5** (100.0) [2.9]

(Contd.)

(Contd. Table 7.11)

Households Monthly Income (Rs.)	*Travel time per head per household*				*Total*
	Less than 10	*10-20*	*20-30*	*30 and above*	
10000 - 12500		**1** (50.0) [2.4]	**1** (50.0) [2.2]		**2** (100.0) [1.1]
12500 - 15000	**1** (100.0) [2.2]				**1** (100.0) [.6]
15000 and above	**1** (33.3) [2.2]		**1** (33.3) [2.2]	**1** (33.3) [2.3]	**3** (100.0) [1.7]
Total	**45** (25.7) [100.0]	**41** (23.4) [100.0]	**45** (25.7) [100.0]	**44** (25.1) [100.0]	**175** (100.0) [100.0]

Source: Survey Data, 2005.
Note : Figure in the parenthesis indicates row percentage.
Figure in the brackets indicates column percentage.

Table 7.12. Distribution of Monthly Household Income according to the Waiting Time

Monthly Household Income (Rs.)	*Total Waiting Time per head per Household (in hours)*				*Total*
	Below 1.30	*2-2.30*	*2.30-3*	*3.30 and above*	
Below 2500	9 [36.0]	9 [37.5]	33 [44.0]	7 [33.3]	58 [40.0]
2500 - 5000	8 [32.0]	12 [50.0]	28 [37.3]	8 [38.1]	56 [38.6]
5000 - 7500	6 [24.0]	2 [8.3]	8 [10.7]	4 [19.0]	20 [13.8]
7500 - 10000	1 [4.0]		3 [4.0]	1 [4.8]	5 [3.4]

(Contd.)

(Contd. Table 7.12)

Monthly Household Income (Rs.)	*Total Waiting Time per head per Household (in hours)*				*Total*
	Below 1.30	*2-2.30*	*2.30-3*	*3.30 and above*	
10000 - 12500		1 [4.2]	1 [1.3]		2 [1.4]
12500 - 15000				1 [4.8]	1 [.7]
15000 and above	1 [4.0]		2 [2.7]		3 [2.1]
Total	25 [100.0]	24 [100.0]	75 [100.0]	21 [100.0]	145 [100.0]

Source: Survey Data, 2005.
Note : Figure in the brackets indicates column percentage.

Therefore, almost equal number of households is found in different classes of travel time. But they are scattered in different classes of household income.

It is clear that only 6.3 per cent of the households are found in the income groups of Rs.7500 and above. However, most of them are found in the classes of lower distance travel.

The general tendency is that the lower income households are found in the classes of more travel time per head. Table 7.12 shows the distribution of monthly household income according to the waiting time.

It is found that 51.7 per cent of the households wait as much as two and half to three hours in the queue to fetch water. The percentages of households waiting for more than three and half-hours for collecting water comprise only 14.5 per cent. Moreover, majority of higher income households are found in the category of lower waiting time.

The in-depth study of water supply unreliability of the households having no in-house water connection reveals that the poor income households face extreme level of unreliability problem due to unpredictability of water supply, low level of supply and long duration of time in collecting water. However, the poor income households find it difficult to adopt compensatory strategies in their effort to reduce the problem of water supply unreliability.

SECTION-B

Compensatory Strategies

As the households face irregular and insufficient water supply, they try to resolve these difficulties by resorting to one or more compensatory strategies. Affluent households are likely to adopt efficient compensatory strategies, while the selection of such strategies by the poor households is from various alternatives, depending upon the level of household's income and the extent of water supply unreliability. The poor make less fixed investment in compensatory strategies because of the real operation and maintenance costs.

Types of Compensatory Strategies

In Chennai City, households adopt one or more than one compensatory strategies, especially to address the problems of water storage and water pressure. The different types of compensatory strategies adopted by the sample households are listed in Table 7.13.

More than three-fourths of the households (75.1per cent) store water in *Kodam* and buckets. Poor households take to this practice as a measure to reduce the problems of unpredictability of water supply. As many as 27.3 per cent households have dug bore wells and they have built complementary overhead tanks in order to store water, especially for bathing and washing purposes. Due to the low

Table.7.13. Types of Strategies used by Households

Types of Strategies	*Location*		*Total*	*Percentage of households*
	North Chennai	*South Chennai*		
Underground Tank (UGT) and Overhead Tank (OHT)	14	35	49	10.8
Kodam (pots) and Buckets	192	146	338	75.1
Dug well	8	2	10	2.2
Own Bore well and Overhead Tank	44	79	123	27.3
Shifting of Residence	8	10	18	4.0
Water Treating	119	151	270	60.0

Source: Survey Data, 2005.

pressure at the tap, a majority of households (47.7 per cent) install hand pumps to draw water.

It is reported that 4 per cent of the households shifted their residence due to shortage of water. Over 10 per cent of the households adopted compensatory strategies such as construction of underground tank to store Metro water and then lift it to the overhead tanks. It is interesting to note that 60 per cent of the households treat water before cooking and drinking.

Almost all the sample households reorganized their water storage strategies in accordance with the extent of water supply. Moreover, three out of every five households adopted treatment of water. One in four households used the sub-soil water to compensate for the non-availability of water for supplementary water needs. Poor people store water in *kodam,* which occupy a larger space of their residence. It is found that 4 per cent of the sample households shifted their residences citing the reason of non-availability of water. Storage strategy is a temporary measure in dealing with insufficient and unpredictable water supply, which is being practiced by the households as a dominant strategy.

Storing Water in *Kodam* and Buckets

Storing of water in *kodam,* buckets and in some other types of vessels is a widely prevalent strategy practiced by 75.1 per cent of the sample households in Chennai City. This kind of water storage strategy is mainly used for cooking and drinking water by a majority of the households. The households in North Chennai (85.3 per cent) rely more on this strategy, when compared to the 64.8 per cent households in South Chennai. It is interesting to note that the poor households adopt this strategy even to store water for their supplementary needs like bathing and cleaning. Some middle income and higher income households purchase water in bottles and in plastic containers for the purposes of drinking and cooking.

Figure.7.4. Households Income and Storing Measures

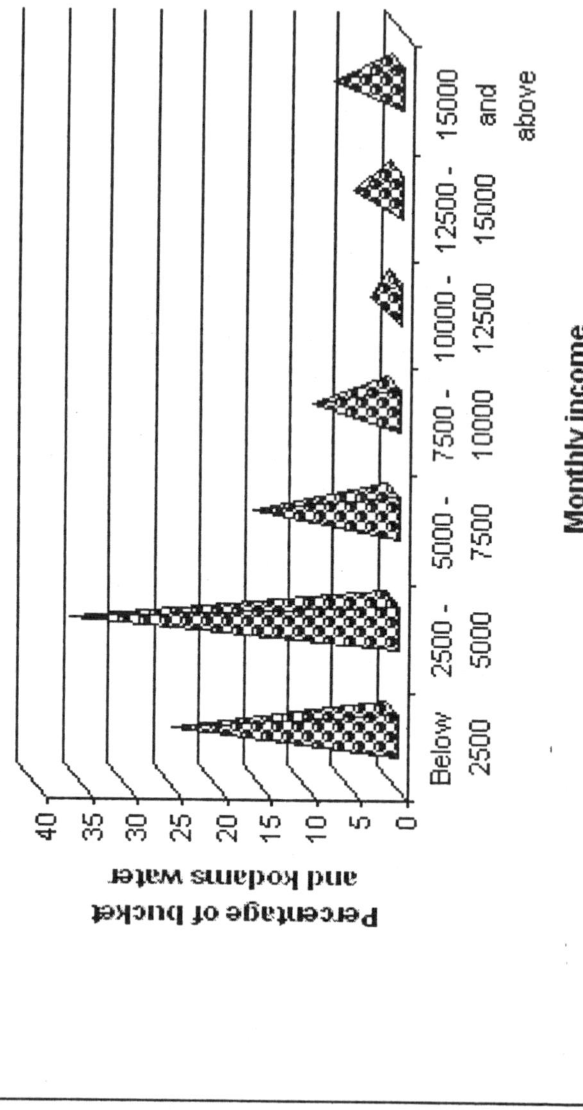

It is perceived that storing of water in *kodams* and buckets is the general practice of the poor income households. Hence, the monthly incomes of the households are compared with the practice of storing water in *kodams* and buckets. Table 7.14 reveals that over three-fifths of the 338 households (60.7 per cent) using *kodams* and buckets to store water earn an income of less than Rs.5000 per month.

Table 7.14. Households Monthly Income and Storing of Water in *Kodams* and Buckets

Monthly Income (Rs.)	*Bucket and Kodams water*	*Percentage*
Below 2500	83	24.6
2500 - 5000	122	36.1
5000 - 7500	53	15.7
7500 - 10000	31	9.2
10000 - 12500	9	2.7
12500 - 15000	16	4.7
15000 and above	24	7.1
Total	338	100.0

Source: Survey Data, 2005.

However, it is inferred that the poor households largely practice storing water in *kodam* and buckets, but the not so poor households are also practicing it as well though the latter households shares a lesser percentage in the respective total when compared to the former households.

Null Hypothesis (H_o):

Lower income households have no inefficient compensatory strategies.

Alternative Hypothesis (H_a):

Lower income households have inefficient compensatory strategies.

The Karl Pearson's correlation coefficient is worked out in order to assess the degree of relationship between monthly household income and the efficiency of the alternative strategies to metro water adopted by the households. The Karl Pearson's correlation coefficient value is estimated as -0.428 and significant at 0.01 levels (2 tailed). It shows that there is a relatively high degree of negative correlation between monthly household income and number of households storing water is *kodam* and bucket. It means that at high monthly household income levels less number of households stored water in *kodams* and buckets. Thus the r value rejected the null hypothesis and accepted the alternative hypothesis.

Storage of Metro Water in Underground Tanks and Overhead Tanks

Underground tanks known otherwise as '*sumps*' and overhead tanks are the two complementary storage devices connected to the distribution network, which allow a greater margin of independence when compared to the storage of water in *kodams* and buckets. Usually the underground and overhead tanks are connected with electric booster pumps. Metro water is stored in underground sumps during its time of supply and then pumped to overhead tanks whether big or small. Some households install low powered electric motor pumps illegally on the water line itself and directly connect it to the overhead tanks. It leads to water contamination and reduces water pressure in the distribution network. Households resort to this strategy when they face acute water supply unreliability and have sufficient income to make necessary investment in constructing underground and overhead water tanks.

It appears that installing underground and overhead tanks by the higher income households are the precautionary measures taken by households to meet acute water supply unreliability. It is observed that some of the households use the underground and overhead tanks to distribute the purchased private water, supplied by tanker lorries when there is no metro water supply. Some well-to-do households have separate underground and overhead tanks for storing Metro water. It is revealed that only 10.9 per cent (49 out of 450) of the households have underground and overhead tanks to store Metro water. Nearly two thirds (65.3 per cent) of the households possessing underground water sumps are in the highest monthly income bracket of Rs.15000 and above. The capacity of underground water sumps also varies directly with the size of monthly income of the households. It is found that all the households having underground water sumps also possess overhead tanks but their availability varies between North Chennai and South Chennai. It is found that in North Chennai only 28.6 per cent households have both underground and overhead tanks. The capacity of their overhead tanks ranges from a minimum of 6000 liters to the maximum of 10,000 litres. In South Chennai, more than half (51.4 per cent) of the households possess overhead tanks with capacity of 8000 litres. The capacity of tanks of the other households varies from 3000 litres to 12,000 litres.

Dug Well Strategy

Dug well strategy is one of the oldest strategies to meet water supply unreliability. Dug wells generally supplement the water needs for bath and washing. Out of the 450 households, only 2.5 per cent (10 households) have their own dug wells. It is a very expensive strategy, requiring space in front, or at the backside of the houses. Out of 10 dug wells, 8 (80 percent) are concentrated in North Chennai. The average width of a dug well is 3 meters and the average depth is 50 feet.

Bore Well Strategy

Bore wells are the common strategy used by the households in Chennai City to tide over the unreliable water supply. By far, installing the bore wells is very expensive. In most cases, the bores are connected with electric motor pumps. However, a few households have attached manual hand pumps. It is found that 27.3 per cent (123 out of 450) of the surveyed households have installed bore wells. Over three fifths (64.2

per cent) of the bore wells are in South Chennai. To a large extent, the scarcity of water in Chennai City is compensated by the availability of ground water. Over-dependence of households on ground water accompanied by population pressure has caused a rapid decline in the ground water table. Hence, they have to incur additional costs to dig deeper bore wells. Approximately, 91.1 per cent of the bore wells have a width radius of 5mm and 34.1 per cent have a depth of 75 ft. It is interesting to note that the average depth of bore wells is estimated to be 120 ft.

The overhead tank is an important strategy to store bore-well water. The overhead tank is usually built on the terrace of the building at the time of house construction. Almost all the households, which have bore-wells, have the facility of overhead tanks. More than 30 per cent of the surveyed households (123) dug bore wells in order to meet the water unreliability problem. South Chennai has more number of bore wells and overhead tanks when compared to North Chennai. Nearly two-thirds (64.3 per cent) of the bore wells and the associated overhead tanks are in South Chennai. It is believed that rich people mainly use them.

Contingent Strategies

As a reaction to water supply unreliability, households reduce water consumption even for the most urgent needs. Hence, unreliability is more of a reaction to a given situation than a matter of choice. In Chennai City, water supply unreliability impacts upon the day-to-day activities of the surveyed households. Usually households have to organize their daily chores of life in comparison to the volume of availability of water. It could be seen in the Chennai City, especially for those households relying on street water sources, to engage the entire family to be most active during the hours of water availability. The problems become cumulative if there is unpredictability in water supply. The members of the households, who spend more time in fetching water in the mid-night, have to undergo severe strain in the daytime, while they perform their normal duties in their work places. The diverse physical and psychological impact, which the unreliability of water lends to include inability to send the children to school and job-goers in time, going out without taking bath, tensions like waiting in the queue for long hours, family conflicts, fatigue due to carrying buckets and pots full of water, etc. For meeting the problems arising out of reorganizing water supply unreliability, the households invest in alternative strategies, even though such investments are loaded with higher opportunity costs.

Another alternative strategy used by the households is shifting their residences due to unreliable nature of water supply in the present area. Wherever water is not available, households usually shift to the areas where there is adequate water supply. It is reported that in one year prior to the date of the Survey, 4 per cent of the sample households shifted their residences for the simple reason that water supply in their place of residence has been unreliable. Such a feature is almost equally represented both in North Chennai and South Chennai.

Qualitative Strategy

The households in Chennai City face not only inadequate quantity of water supply but also the problems of poor quality of water. Appropriate water treatment strategy has been a regular strategy widely adopted by most of the households. A majority of the households treat water for the specific reason that they want to get rid of some water-borne diseases like jaundice, typhoid, etc. Hence the households were asked about their practice of treating of water everyday. The results are portrayed in Table 7.15.

It is reported that 270 (60 per cent) of the 450 sample households treat water for drinking purpose everyday. But this is being practiced by more number of households in South Chennai, compared to the North Chennai, i.e., 67.1 per cent in

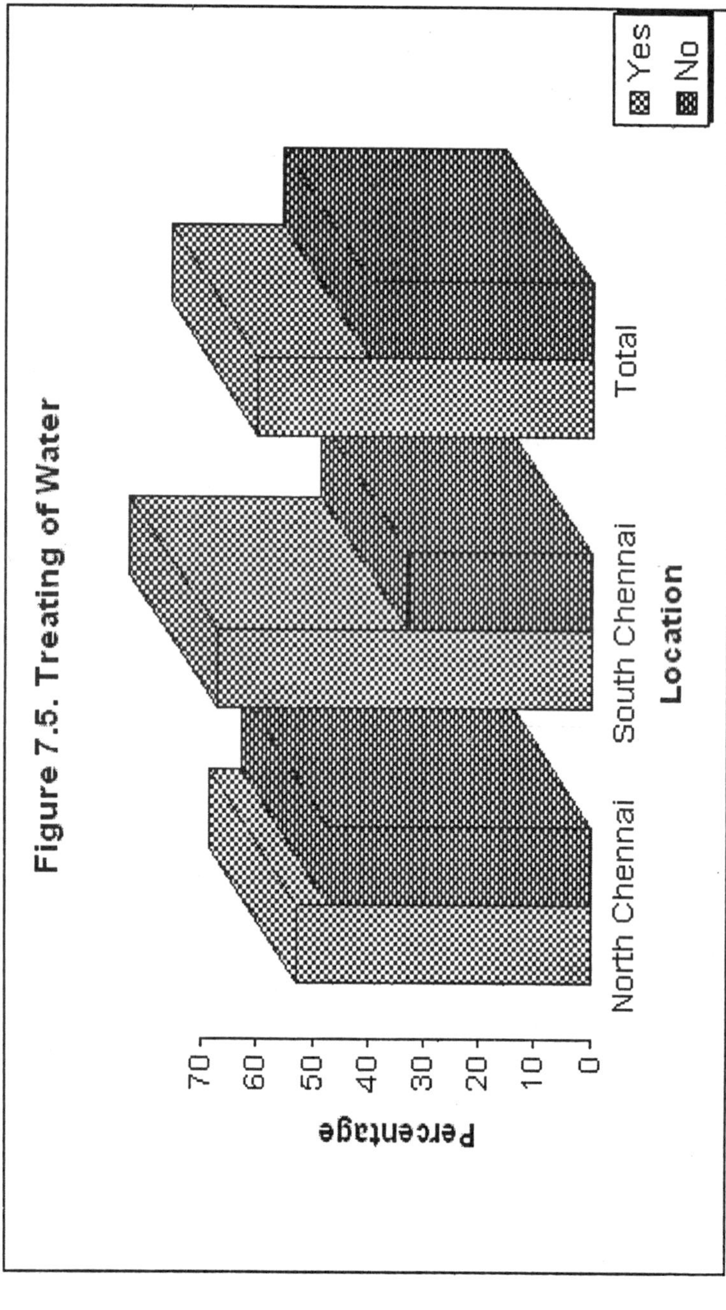

Figure 7.5. Treating of Water

Table 7.15. Treating of Water

Treating water every day	*Location*		*Total*
	North Chennai	*South Chennai*	
Yes	119 (44.1) [52.9]	151 (55.9) [67.1]	270 (100.0) [60.0]
No	106 (58.9) [47.1]	74 (41.1) [32.9]	180 (100.0) [40.0]
Total	225 (50.0) [100.0]	225 (50.0) [100.0]	450 (100.0) [100.0]

Source: Survey Data, 2005.
Note : Figure in the parenthesis indicates row percentage.
Figure in the brackets indicates column percentage.

South Chennai as compared to 52.9 per cent in North Chennai. The practice of treating water for drinking purpose and the respective household's income create a base for comparative analysis. The results are furnished in Table 7.16.

It is found that water treating practice is widely prevalent among the lower income households because of their dependence

Table 7.16. Household's Monthly Income and the Number of Households Treat Water Every Day for Consumption

Households Monthly Income (Rs.)	*Number of households Treat water every day*	*Percentage*
Below 2500	29	10.7
2500 - 5000	70	25.9
5000 - 7500	43	15.9
7500 - 10000	39	14.4
10000 - 12500	8	3.0
12500 - 15000	24	8.9
15000 and above	57	21.1
Total	270	100.0

Source: Survey Data, 2005.

on metro water for drinking purpose. However, households in the higher income groups rely more on packaged water for drinking purpose where the question of treating of water does not arise.

SECTION-C

Cost of Unreliability

Compensatory strategies adopted by the households are the reflections of unreliability of water supply; however, it does not quantify the socio-economic costs of such strategies. In this section an attempt is made to estimate the costs of unreliability. The total cost of unreliability can be classified into (a) the costs of avoiding unreliability and (b) the costs of undergoing such unreliability. The cost of *'avoiding'* unreliability includes both operating cost and the cost of new investment in various strategies. Broadly speaking, unreliability is not only borne directly by the households but also indirectly by the society as a whole. The costs of reducing unreliability and the cost of water treatment or investment in alternative strategies are easy to estimate as compared to the costs arising out of water borne diseases. If unreliability has an effect on the quality of water, quantification of costs resulting there from are very difficult.

The cost of reducing unreliability comprises both economic and opportunity costs. The estimation of economic cost comprises investment in appliances and their operating and maintenance costs. The time opportunity costs are the monetary value of time lost in the processes of collecting water, standing in the queue, filling the buckets, and *kodams*, hand pumping, the costs of water filters or water purifiers, etc., the costs of boiling water and other water treatment costs. The medical expenses and loss of income due to sickness owing to poor quality of water come under the category of costs of water borne diseases.

Table 7.17. Descriptive Statistics of Total Unreliability Water Supply Cost (Rs.)

Average annual cost of unreliability components	*N*	*Mean (Rs.)*	*S. Deviation*
Underground tank (sump)	49	223.2993	78.96501
Overhead tank	49	132.9932	32.54089
Motor to lift Metro water	49	346.9388	49.49671
Electricity expenditure to Metro water	49	130.4082	45.22901
Hand pumps	217	51.5276	6.79216
Kodams (pots)	18	217.3333	156.24942
Buckets	400	526.4875	178.48527
Barrels	7	107.6190	48.57084
Digging dug wells	10	171.6667	42.34631
Repair and maintenance of dug wells	10	320.0000	88.81942
Digging bore wells	123	1875.6098	637.66241
Overhead tanks to lift bore water	123	165.5827	61.75521
Motors to lift bore water	123	662.3306	247.02084

(Contd.)

(Contd. Table 6.17)

Average annual cost of unreliability components	*N*	*Mean (Rs.)*	*S. Deviation*
Electricity expenditure to bore water	123	530.6341	238.87871
Repair and maintenance of bore well	123	630.2927	451.00149
Relocating residences	17	350.0000	186.24581
Cost of time spent in water collection	175	14425.7143	3433.44712
Energy used to boil water	238	546.6050	211.89760
Filter device costs	8	400.0000	106.90450
Maintaining filter costs	8	336.2500	226.77475
Aqua guard costs	23	650.0000	83.93721
Energy to use purify water	23	357.8261	292.80705
Water bone diseases treatment	69	464.3478	197.88944
Metro water supply in lorry	17	22.0294	60.37655
Purchased water	47	359.7872	693.45618
Total Unreliability Cost	450	7838.9345	6866.79753

Cost calculations are based on compiled price data and survey details.

The costs of unreliability of water supply for each household are based on the Survey data and on the secondary data regarding the prices of various appliances used and the maintenance and depreciation costs. For this purpose, the capital opportunity costs are estimated on an annual basis, depending upon the approximate life of each appliance and the total investments made. The time opportunity costs are also assessed on annual basis. This refers to the wage loss, if alternative employment is available to them during the lost period in water collection. The costs of water borne diseases are the costs incurred by a household for medical expenses and the consequent wage loss due to sickness in the last one year. The average cost of unreliability is measured on annual basis, which is inclusive of capital opportunity costs, time opportunity costs and the cost of treatment against water-borne diseases. The descriptive statistics of the various components of unreliability costs are shown in Table 7.17.

The average cost of unreliability per household is estimated as Rs.7838.93 per year. But the households with in-house water connection pay Rs.600 per year as water tariff i.e., 13 times less than the estimated *'unreliability cost'*. Among all the unreliability cost components, the average cost of digging bore well stands first, at Rs.1875.61 with a comparatively lesser standard deviation of 637.6624. The biggest component of unreliability cost is the average cost of time spent for water collection, which is estimated as Rs.14,425.71 per year. Most of such households are poor who use the common source for water fetching.

In order to understand the impact of unreliability on different sections of the households, the average annual unreliability cost has been compared with some of the determinants like size of the households, monthly household income, total floor area of the house, educational standards of each head of the households, source of water and per capita water consumption.

Table 7.18. Average Unreliability Cost and the Size of the Household

Average unreliability cost (Rs.)	*Size of the Household (Numbers)*					*Total*
	Below 2	*2 - 4*	*4 - 6*	*6 - 8*	*8 and above*	
Below 2500	22 [44.0]	70 [32.7]	54 [35.1]	6 [21.4]	2 [50.0]	154 [34.2]
2500-5000	9 [18.0]	31 [14.5]	14 [9.1]	3 [10.7]		57 [12.7]
5000-7500	7 [14.0]	26 [12.1]	15 [9.7]	7 [25.0]	1 [25.0]	56 [12.4]
7500-10000		4 [1.9]	3 [1.9]			7 [1.6]
10000-12500	7 [14.0]	29 [13.6]	14 [9.1]			50 [11.1]
15000-17500	4 [8.0]	39 [18.2]	31 [20.1]	8 [28.6]		82 [18.2]
17500 and above	1 [2.0]	15 [7.0]	23 [14.9]	4 [14.3]	1 [25.0]	44 [9.8]
Total	50 [100.0]	214 [100.0]	154 [100.0]	28 [100.0]	4 [100.0]	450 [100.0]

Source: Survey Data, 2005.
Note : Figure in the brackets indicates column percentage.

Average Unreliability Cost and the Size of the Household

The size of the household is one of the important factors influencing the average unreliability cost for a household. Table.7.18 shows the cross tabulation of average unreliability costs estimated per house and the size of the household. It is true that the average unreliability cost hides considerable differences among the households. Different households adopt different alternative strategies depicting huge variations in capital opportunity costs.

Time opportunity costs are the consequence of the households drawing water from outside their residences and thus spending considerable time for fetching water. Many of the households do not recognize the opportunity costs of time spent for water collection.

It is evident that 47.6 per cent of the households have a family size ranging from 2-4 members, of whom nearly half (47.2 per cent) incur an average annual unreliability cost of less than Rs.5000. But, when the family size increases, relatively less number of households is found in the average unreliability classes of less than Rs.5000. The Karl Pearson's correlation coefficient between the averages of unreliability cost per household and the size of the household is computed as 0.138 at one per cent significance level. It confirms a low degree of positive correlation between the average unreliability cost and size of the household.

Average Unreliability Cost and Monthly Household Income

Average unreliability cost is distributed according to the monthly household income in Table 7.19.

The best indicator of *'unreliability'* is the proportion of income spent on compensatory strategies. But it may not reflect

Table 7.19. Average Unreliability Cost and Monthly Household Income

Average unreliability cost (Rs.)	*Monthly Household Income*							
	Below 2500	*2500 - 5000*	*5000 - 7500*	*7500 - 10000*	*10000 - 12500*	*12500 - 15000*	*15000 and above*	*Total*
Below 2500	17 [19.5]	42 [32.1]	27 [42.9]	19 [40.4]	6 [50.0]	17 [51.5]	26 [33.8]	154 [34.2]
2500-5000	2 [2.3]	9 [6.9]	5 [7.9]	14 [29.8]	3 25.0]	6 [18.2]	18 [23.4]	57 [12.7]
5000-7500	3 [3.4]	6 [4.6]	5 [7.9]	8 [17.0]	1 [8.3]	7 [21.2]	26 [33.8]	56 [12.4]
7500-10000			1 1.6]	1 [2.1]		1 [3.0]	4 [5.2]	7 [1.6]
10000-12500	18 [20.7]	21 [16.0]	8 [12.7]		1 [8.3]	1 [3.0]	1 [1.3]	50 [11.1]
15000-17500	36 [41.4]	33 [25.2]	8 [12.7]	4 [8.5]	1 [8.3]			82 [18.2]
17500 and above	11 [12.6]	20 [15.3]	9 [14.3]	1 [2.1]		1 [3.0]	2 [2.6]	44 [9.8]
Total	87 [100.0]	131 [100.0]	63 [100.0]	47 [100.0]	12 [100.0]	33 [100.0]	77 [100.0]	450 [100.0]

Source: Survey Data, 2005.
Note : Figure in the brackets indicates column percentage.

the time opportunity costs. Hence, the average unreliability cost is compared with the respective average monthly household incomes. It can be seen that a majority of the 34.2 per cent of the households belonging to the lowest unreliability class of below Rs.2500. The general trend is that the percentage of households in the higher unreliability classes increases with a decrease in household income.

The highest monthly household income class of Rs.15, 000 and three-fourth (77.5 per cent) of the households in the highest monthly income class of Rs.15,000 and above are found in the average unreliability cost classes of less than Rs.7500. While analysing the highest average unreliability cost of Rs.17, 500 and above, it can be seen that 91 per cent of them earn an income of less than Rs.7500 per month.

Null Hypothesis (H_0):

Household income is not inversely related to average household annual unreliability cost.

Alternative Hypothesis (H_a):

Household income is inversely related to average household annual unreliability cost.

In order to estimate the relationship between average unreliability cost and the monthly household income, Karl Pearson's Correlation Coefficient is estimated. The correlation coefficient of -0.33 at 0 per cent significance level shows that the average unreliability cost decreases with an increase in monthly household income. It means that the null hypothesis is rejected and alternative hypothesis is accepted.

Average Unreliability Cost and Literacy Standard

Table 7.20 shows the comparison between average unreliability costs and the literacy standard of the heads of the households.

Table 7.20. Average Unreliability Cost and Educational Standard

Average Unreliability cost (Rs.)	*Educational standard*				
	Illiterate	*Primary*	*Middle*	*High school*	*Higher secondary / PUC*
Below 2500	19 [20.0]	6 [31.6]	26 [32.9]	64 [46.0]	15 [37.5]
2500-5000	7 [7.4]		8 [10.1]	11 [7.9]	9 [22.5]
5000-7500	6 [6.3]	1 [5.3]	8 [10.1]	16 [11.5]	5 [12.5]
7500-10000			1 [1.3]	3 [2.2]	1 [2.5]
10000-12500	15 [15.8]	3 [15.8]	11 [13.9]	13 [9.4]	3 [7.5]
15000-17500	30 [31.6]	7 [36.8]	16 [20.3]	20 [14.4]	5 [12.5]
17500 and above	18 [18.9]	2 [10.5]	9 [11.4]	12 [8.6]	2 [5.0]
Total	95 [100.0]	19 [100.0]	79 [100.0]	139 [100.0]	40 [100.0]

(Contd.)

(Contd. Table 7.20)

Average Unreliability cost (Rs.)	*Educational standard*				*Total*
	Graduate	*Post-graduate*	*Professional*	*Others*	
Below 2500	18 [38.3]	6 [27.3]			154 [34.2]
2500-5000	9 [19.1]	8 [36.4]	4 [66.7]	1 [33.3]	57 [12.7]
5000-7500	12 [25.5]	6 [27.3]	2 [33.3]		56 [12.4]
7500-10000	1 [2.1]	1 [4.5]			7 [1.6]
10000-12500	4 [8.5]			1 [33.3]	50 [11.1]
15000-17500	3 [6.4]	1 [4.5]			82 [18.2]
17500 and above				1 [33.3]	44 [9.8]
Total	47 [100.0]	22 [100.0]	6 [100.0]	3 [100.0]	450 [100.0]

Source: Survey Data, 2005.
Note : Figure in the brackets indicates column percentage.

Table 7.21. Average Unreliability Cost and Total Floor Area of the House

Average unreliability cost (Rs.)	*Total Floor Area of the House (sq.ft.)* Below 300	300-600	600-900	900-1200	1200-1500	1500 and above	Total
Below 2500	3 [14.3]	34 [23.9]	55 [44.4]	27 [37.0]	28 [47.5]	7 [22.6]	154 [34.2]
2500-5000		7 [4.9]	19 [15.3]	15 [20.5]	10 [16.9]	6 [19.4]	57 [12.7]
5000-7500	1 [4.8]	3 [2.1]	15 [12.1]	13 [17.8]	13 [22.0]	11 [35.5]	56 [12.4]
7500-10000			3 [2.4]	2 [2.7]	1 [1.7]	1 [3.2]	7 [1.6]
10000-12500	5 [23.8]	35 [24.6]	5 [4.0]	2 [2.7]	2 [3.4]	1 [3.2]	50 [11.1]
15000-17500	9 [42.9]	47 [33.1]	15 [12.1]	7 [9.6]	4 [6.8]		82 [18.2]
17500 and above	3 [14.3]	16 [11.3]	12 [9.7]	7 [9.6]	1 [1.7]	5 [16.1]	44 [9.8]
Total	21 [100.0]	142 [100.0]	124 [100.0]	73 [100.0]	59 [100.0]	31 [100.0]	450 [100.0]

Source: Survey Data, 2005.
Note : Figure in the brackets indicates column percentage.

The illiterates account for 21.1 per cent of the respondents, of whom more than half represents the highest annual unreliability cost classes of Rs.15,000-17,000 to Rs.17,000 and above.

This trend is visible when moving towards higher standards of literacy. Therefore, it is hypothesised that there is no association between literacy levels of the respondents and the average unreliability cost. Since the calculated value of $\div^2$ is more than the table value, connote the rejection of null hypothesis and therefore the alternative hypothesis is accepted.

Average Unreliability Cost and the Total Floor Area of the House

The average unreliability cost per household is compared with the total floor area of the house. The details are furnished in Table 7.21.

Null Hypothesis (H_0):

Area of the house is not inversely related to average household annual unreliability cost.

Alternative Hypothesis (H_a):

Area of the house is inversely related to average household annual unreliability cost.

The data demonstrates that unreliability is high in the case of households having lesser floor area. In order to find the relationship between the floor area of the house and average unreliability cost, Karl Pearson's Correlation coefficient is worked out. The value of Correlation coefficient is -313 at zero level of significance indicates that the higher the area of the house, lower will be the level of average unreliability cost.

Average Unreliability Cost and Source of Water

The classification of households according to the nature of water supply source and the average unreliability cost per households are shown in Table 7.22.

Table 7.22. Average Unreliability Cost and Source of Water

Average unreliability cost (Rs.)	*Source of water*		*Total*
	In-house metro water connection	*Outside metro water source*	
Below 2500	153 (99.4) [57.5]	1 (.6) [.6]	154 (100.0) [35.1]
2500-5000	52 (100.0) [19.5]		52 (100.0) [11.8]
5000-7500	50 (100.0) [18.8]		50 (100.0) [11.4]
7500-10000	7 (100.0) [2.6]		7 (100.0) [1.6]

(Contd.)

(Contd. Table 6.22)

Average unreliability cost (Rs.)	*Source of water*		*Total*
	In-house metro water connection	*Outside metro water source*	
10000-12500	1 (2.0) [.4]	49 (98.0) [28.3]	50 (100.0) [11.4]
15000-17500	1 (1.2) [.4]	81 (98.8) [46.8]	82 (100.0) [18.7]
17500 and above	2 (4.5) [.8]	42 (95.5) [24.3]	44 (100.0) [10.0]
Total	266 (60.6) [100.0]	173 (39.4) [100.0]	439 (100.0) [100.0]

Source: Survey Data, 2005.
Note : Figure in the parenthesis indicates row percentage.
Figure in the brackets indicates column percentage.

It gives a clear picture that 95.8 per cent of the households with in-house water connection have an average unreliability cost of less than Rs.7, 500 per year whereas 99.4 per cent of the households without in-house water connection incur an average unreliability cost of Rs.10, 000 and above. Hence it is quite clear that the average unreliability cost of households collecting water from outside the house is heavy, when compared to the households with in-house water connection.

The Design of Contingent Valuation Method (CVM)

Referendum elicitation procedure is a common technique used in the CVM. In this method, closed-ended question is formulated to get yes or no answer. Among various elicitation formats, the single-bounded dichotomous choice questions were chosen to obtain households WTP for existing and improved water supply. Of the 450 sample households, 264 (58.7 per cent) have in-house water connection and 38.8 per cent (175) totally depend on common water distribution points while 2.5 per cent households (11) do not depend on public water supply sources since they have the water source of their own. Hence, the willingness to pay of all the three categories of households may not be the same. In order to maintain affordability criterion, WTP of households with in-house water connection and households without in-house water connection have been treated separately. The households even in the same category have different WTP for the existing water supply and improved water supply. Thus, totally four contingent valuation methods have been used. They are as follows.

1. Households connected to in-house water supply.
(1) WTP for the existing water supply
(2) WTP for the improved water supply

2. Households not connected to in-house water supply
(1) WTP for the existing water supply
(2) WTP for the improved water supply

Table.7.23. Package for Improved In-House Water Supply

Outcome of change	*Alternatives*				
	A	*B*	*C*	*D*	*E*
Reliability (Days)	3 days	4 days	4 days	7 days	7 days
Hours per day	6 hrs	6 hrs	8 hrs	10 hrs	12 hrs
Pressure	Low	High	Medium	Medium	High
Quality	Medium	Medium	Medium	High	High
Bid amounts	90	100	110	120	130

The bidding method of contingent valuation is used for both existing water supply and improved water supply. The bid amounts are finalized after carefully examining the socio-economic characteristics of the households. The bid amount for maintaining the existing water supply starts from Rs.50 to Rs.90 per month containing a total of five bid amounts having an interval of Rs.10. The WTP for improved water supply is based on a package containing various characteristics. These characteristics are given in Table 7.23.

It is found that the ability to pay of the households depending on public water supply is lesser than that of the households connected to the in-house water supply. Therefore the starting point of bid amount of both existing and improved water supply for the households depending on public water source is lesser. The bid amount for the existing water supply starts from Rs.10 to Rs.30 per month containing a total of five bid amounts having an interval of Rs.5 each. The improved package offers various characteristics. These are given as:

1. The distance between residence and stand post not to exceed 100 meters.
2. Totally ten households will be entitled to collect water from a single tap.
3. Water will be supplied everyday two times in the morning and evening for two hours respectively. (i.e., 6 A.M to 8 A.M and 5 P.M to 7 P.M).

4. Allowed to collect 15 *kodams* of 15 litres per day.
5. Monthly tariff will be collected first week of every month.
6. Water will be supplied of medium pressure and quality.

   Based on the above package the bid amount for improved water services ranges from Rs.50 to Rs.90 per month.

A household WTP for an improved water supply service would be a function of the proposed changes in the attributes of services and other factors that influence the households' valuation. It is hypothesized that the probability of responding 'yes' to an existing and proposed water supply is a function of household characteristics perception of the household related to the problems of water supply and the coping strategies adopted by them. The household characteristics include household size (SIZ), sex of the respondent (SEX), educational standard of the respondent (EDN) and monthly household income (MET). In reality the 'yes' response depends on how large the bid amount, which in turn depends upon the households' affordability. The perceptions of the household about water problem related to water pressure and water outage. The coping strategies related to the ways of meeting the water supply unreliability.

It is found that out of the 264 households have in-house water connections only 36 per cent (95 households) expressed their willingness to pay for the existing water supply. The mean value of water bill per month is estimated as Rs.30.89, while the mean bid amount is turned as Rs.46.34, which shows an increase of 50 per cent. However, 67 per cent of (176 out of 264) households having in-house water connections expressed their opinion in favour of paying the improved in-house water supply. Since the proposed water supply is reliable, the mean value of the bid amount for the improved water supply appeared as Rs.68 per month, which shows an increase of 121 per cent when compared to the mean value of the existing water bill of Rs.30.67.

At present all the households fetching water from common distribution points get water at free of cost, even though some

Table.7.24. Households Monthly Income and Number of Households WTP

Household Monthly Income	*WTP for In-House Water Supply*		*WTP for Public Water Supply*	
	'Yes' response for Existing water supply	*'Yes' response for Improved water supply*	*'Yes' response for Existing water supply*	*'Yes' response for Improved water supply*
Below 2500	9 (45.00)	12 (60.0)	10 (15.4)	54 (83.1)
2500-5000	16 (29.6)	41 (75.9)	17 (23.6)	56 (75.7)
5000-7500	15(39.5)	26 (68.4)	7 (28.0)	18 (72.0)
7500-10000	13 (33.3)	26 (66.7)	2 (40.0)	3 (60.0)
10000-12500	3 (30.0)	6 (60.0)	1 (50.0)	1 (50.0)
12500-15000	11 (36.7)	21 (70.0)	3 (100.0)	-
15000 and above	28 (38.4)	44 (60.3)	1 (33.3)	2 (66.7)
Total	95 (36.0)	176 (66.7)	41 (23.4)	134 (76.6)

Source: Survey Data, 2005.
Note :Figures in the brackets indicate percentages of household's willingness to pay.

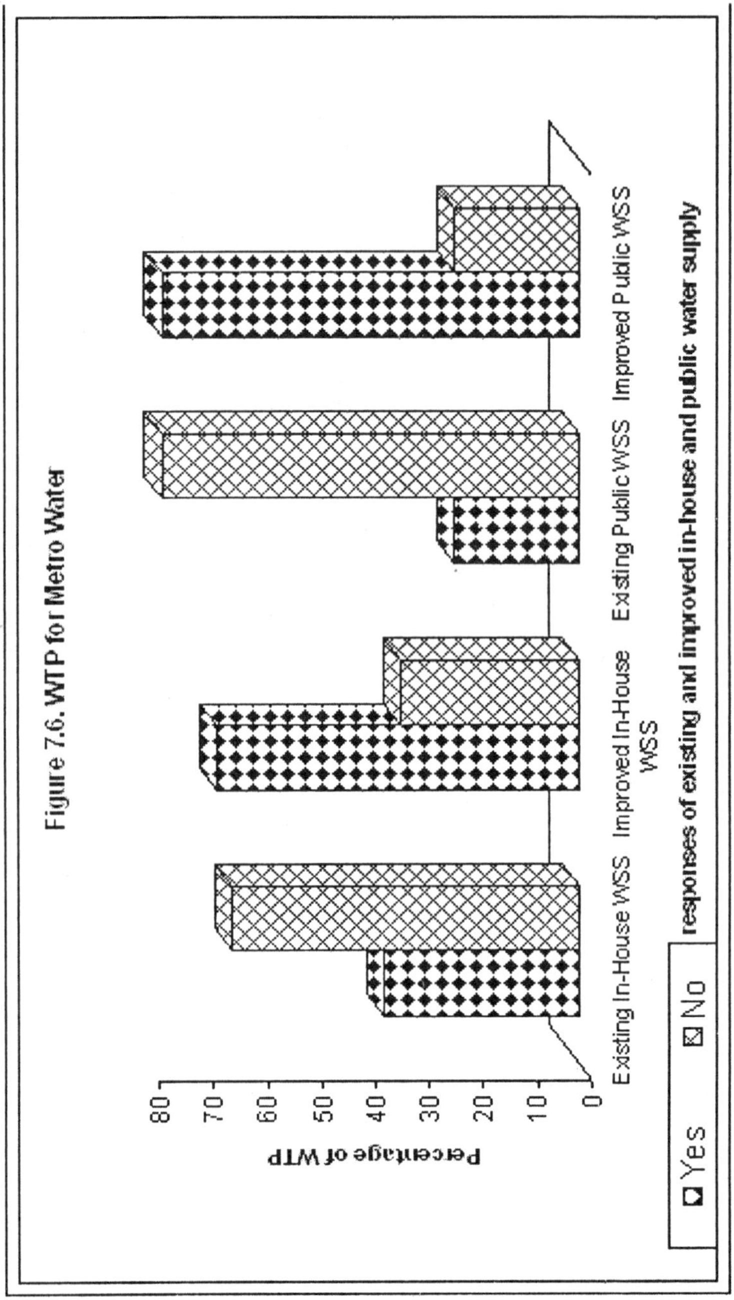

Figure 7.6. WTP for Metro Water

implicit cost is involved in it. In the present model a contingent valuation introducing some bid amount for both existing and a proposed package of improved water supply has been made. To the surprise 23 per cent of the households (41 out of 175) fetching water from the common distribution points WTP some amount for the existing water supply. The mean bid amount value stands at Rs.8.53 per month. Contrary to the general perceptions, 143 of the 175 households (77 per cent) fetching water from common distribution points is WTP some amount for the proposed package of improved water supply. The mean bid amount value for improved water supply is found as Rs.17.58 per month.

In order to assess the responses of willingness to pay for water services in the backlash of their affordability, the monthly household income of the respondents are compared with their willingness to pay for water services. The results are furnished in Table 7.24.

More of the higher income households are willing to pay for the existing in-house water supply. But in all the income classes, majority of the households are willing to pay for an improved in-house water supply. It ranges from 60 per cent to a maximum of 75.9 per cent both are present in the lower income classes. It is interesting to note that the percentage of households willing to pay for the existing public water supply increases when more up in the income ladder. However, the percentage shares of households willing to pay for the proposed improved services in the lower income classes are more when compared to the higher income classes.

Factors Determining Willingness to Pay

The main purpose of the survey is to determine the WTP of households for the existing and hypothetical improved water supply. There are various socio-economic factors predict the WTP. In order to assess the probability of the households paying for water supply, a Logit Multi Regression Model (Binary Choice

Model) is run for the existing and improved water supply of both households connected to in-house water supply and public water supply. Thus four logit models are run one each for a case.

Logit Model

Using logit models does the exploration of whether a household is willing to pay for water supply in Chennai City. This model was chosen because of its ability to deal with a dichotomous dependent variable and various other predictor variables.[2] The model is specified as follows:

$$P_i = E(Y = 1 \mid X_i) = [\ 1/\{1 + e^{-(b0 + bi\text{å}\ Ci)}\}] \qquad (2)$$

Where P_i = is the probability that Y =1, X_i is a set of independent variables explained above and b_i are coefficients to be estimated corresponding to logistic distribution. Taking a natural logarithm of e.q. (2) we obtain

$$L_i = ln\ \{P_i \text{£}(1 - P_i)\} = b_0 + b_i \text{å}\ C_i + e_i \qquad (3)$$

Where L_i, which is called logit, is the log of the odd ratios and is linear in both independent variables and parameters. The estimation method to be used will be maximum likelihood estimator (MLE).

Model-I WTP for Existing Condition of In-house Water Supply

In this model WTP for in-house water supply by the households is expressed as a function of several attributes. The logit model has taken WTP as the dichotomous criterion variable regressed against ten independent variables. The equation to be estimated is:

WTPEIWS = b0 +b1 SEX + b2 AGE + b3 SIZ + b4 EDN +b5 MET + b6WAR + b7 BID + b8 WAC + b9 LOP +b10 TRE.

WTPEIWS	:	Respondents' willingness to pay for existing in-house water supply,
SEX	:	Sex of the respondent (1 for male and 0 for otherwise)
AGE	:	Age of the respondent (years)
SIZ	:	Family size (nos.)
EDN	:	Respondents educational standard (1 for literate and 0 for otherwise)
MET	:	Monthly income of the households
WAR	:	Water price (Rs.)
BID	:	Bid amount (Rs.)
WAC	:	Water colour (1 for Good and 0 for otherwise)
LOP	:	Low pressure (1 for low pressure and 0 for otherwise)
TRE	:	Water treatment (whether the respondent treated water before drinking ;1 for treated and 0 for otherwise)
LOC	:	Location (1 for North Chennai and 0 for Otherwise)

Logit Multi-Regression Results for WTP for Existing In-House Water Supply

Logit regression results for existing supply of in-house water connection revealed that family size (SIZ), educational standard of the respondent (EDN), water price (WAR), low pressure (LOP) and water treatment (TRE) negatively influence the WTP for existing water supply. This is because of the fact that the coefficient of all the above cited predictor variable turned negative (Table 7.25).

However, SEX, AGE, MET, WAR, BID and WAC show the positive coefficients that suggest that when the values of these variable increases, the WTP for existing water supply also increases. Among the regressed predictor variables, water price (WAR) and literacy standard (EDN) of the respondent have the maximum negative marginal effects.

Table.7.25. Logit Multi-Regression Results for WTP for Existing In-House Water Supply

Variable	*Coefficient*	*Marginal effect*	*t-value*	*Prob.*
Constant	-3.507792	-.416897	-3.442	.0006***
SEX	.175412	.208476	.402	.6875
AGE	.120059	.142689	1.158	.2470
SIZ	-.125026	.142689	-1.321	.1866
EDN	-.601901	-.715356	-1.462	.1438
MET	.162447	.193067	.895	.3707
WAR	-.726466	-.863398	-7.703	.0000***
BID	.484311	.575599	7.466	.0000***
WAC	1.445343	.171777	3.820	.0001***
LOP	-.270155	-.321078	-.809	.4184
TRE	-.264993	-.314945	-2.141	.0322**
LOC	.672487	.799245	1.969	.0489

Log Likelihood Value	-181.1010
Chi-Square	101.6805
Significance level	.000
Psudo R^2	0.21
No.of observation:	264

It shows that the odds of the event is decreased significantly by an increase in the water price and literacy standard of the respondent by one unit. This implies that the present quality and quantity of water supply do not encourage the literate respondents to pay for the services.

On the average 1 percent increase in the price (WAR) of existing water supply decreases the marginal probabilities of households using existing water supply approximately 0.86 whereas a 1 per cent increase in bid amount (BID) increases the marginal probability by 0.57. Water colour (WAC) from existing water supply sources increases 1 percent, the marginal probability also increases by 0.17 While treating existing water supply (TRE) leads to a significant decline in marginal probability by almost 0.31. Since the existing water price paid by the households connected to the in-house supply has been selected as the starting bid price of the respondents are not WTP a better bid price, because of the acute unreliability of the present water supply.

Usually size of the household (SIZE) is expected to have positive relationship with the WTP value. When the household size increases higher will be the water requirement and therefore the WTP value would be greater. In the present study, households' size does not have any significant influence but is shows a negative sign on the WTP value. This can also be attributed to the reasons of low water pressure and irregular water supply. It is found that the monthly household income (MET) negatively and insignificantly influences the WTP value. Households with higher income refused to pay water because of the availability of alternative strategies.

The log likelihood value indicates that the amount of variation explained is significantly different from zero. The $\div^2$ value is highly significantly proving the evidence that the regression coefficients are significantly different from zero. The

R^2 value of the model predicts only 21 per cent of the cases are rightly predicted of the selected observations.

Model-II WTP for Improved In-house Water Supply

Influence of the predictor variables on the WTP for improved in-house water supply is shown in the following econometric form. In this model WTP for in-house water supply by the households is expressed as a function of several attributes. The logit model has taken WTP as the dichotomous criterion variable regressed against ten independent variables. The equation to be estimated is:

WTPEIWS = b0 +b1 SEX + b2 AGE + b3 SIZ + b4 EDN +b5 MET + b6WAR + b7 BID + b8 WAC + b9 LOP +b10 TRE.

WTPEIWS	:	Respondents' willingness to pay for existing in-house water supply,
SEX	:	Sex of the respondent (1 for male and 0 for otherwise)
AGE	:	Age of the respondent (years)
SIZ	:	Family size (nos.)
EDN	:	Respondents educational standard (1 for literate and 0 for otherwise)
MET	:	Monthly income of the households
WAR	:	Water price (Rs.)
BID	:	Bid amount (Rs.)
WAC	:	Water colour (1 for Good and 0 for otherwise)
LOP	:	Low pressure (1 for low pressure and 0 for otherwise)
TRE	:	Water Treatment (whether the respondent treated water before drinking (1 for treated and 0 for otherwise)
LOC	:	Location (1 for North Chennai and 0 for otherwise)

Results and Discussion

The predictor variables determining WTP for existing in-house water supply have been the same for WTP for improved

Table.7.26. Logit Multi-Regression Results for WTP for Improved In-House Water Supply

Variable	*Coefficient*	*Marginal effects*	*t-value*	*Prob.*
Constant	-3.15462	-.491425	-2.440	.0147
SEX	.60226	.938197	1.143	.2531
AGE	.70233	.109409	.617	.5374
SIZ	.18130	.282439	1.739	.0820*
EDN	-.98488	-.153422	-2.042	.0412**
MET	-.15748	-.245330	-.701	.4831
WAR	.16655	.166556	1.797	.0723*
BID	.34766	.347668	5.665	.0000***
WAC	.38588	.601127	1.006	.3146
LOP	1.14936	.179047	3.505	.0005***
TRE	.74715	.116391	.750	.4532
LOC	-1.52597	-.237709	-3.677	.0002***

Log Likelihood Value = -146.9537
Chi-Square = 308.4108
Significance level =. 0000000
Psudo R^2 =0.51
No.of observation : 264

in-house water supply. It is to be noticed that the effect of some of the variable included in the model are entirely different from the effects of the same variables in the previous model. For instance the variable SEX, which had no significant influence on the WTP for existing water supply, is influencing the WTP for improved water supply. (See Table 7.26) Similarly WAC, LOP and TRE have very significant influence on the WTP for improved water supply as compared to the existing water. This is because of the fact that the coefficients of these variables are very high.

Though monthly household income affects negatively to the WTP, it does not have any significant influence on WTP as compared to the educational standard of the respondents. The coefficients of the EDN have a significant negative influence on the WTP for improved water supply.

One of the reasons may be that the literate respondents are quite aware of the sensitiveness of democratic governments on water charges. The uncertainty over the implementation of the package may also influence their decision of WTP. Except EDN and MET all other prediction variables selected in the model influence positively to their decision on WTP because they are the indicators of the water supply unreliability.

The marginal effects of the predictor variables revealed that a one percent increase of women as the respondent increases the marginal probability of WTP by 0.94. The other predictor variables having larger marginal effects are water colour (WAC) and bid amount (BID) and household size (SIZ). The result implies that factors related to water supply unreliability influence the decision of the respondent's acceptance for payment. The R^2 value of the model explains that the regressed variables predict 51 per cent of the cases willing to pay. Moreover, the $\div^2$ value if highly significant at zero per cent significance level.

WTP for Public Water Supply

As against the general perception, it is found that the households are willing to pay for water through the public taps as well. Though the households have no formal experience of paying water tariff, they expressed their willingness to pay for public water. However, only a few households came forward to pay for the existing water supply when compared to the households WTP for improved hypothetical package of water supply.

Mode III WTP for Existing Public Water Supply

A brief portrayal of the unreliability problems of households depending on public water source is already made. Each of the factors contributed to the problem of unreliability is studied while examining their effort to meet the required water demand. Then each respondent was asked his willingness to pay for the existing water supply. There are only 23 per cent of the people getting water from public source replied positively. A logit regression model is run to predict the probability of WTP for the existing public water supply.

The econometric expression of the regression equation is given as:

$$WTP_{EPWS} = b_0 + b_1 SEX + b_2 AGE + b_3 SIZ + b_4 EDN + b_5 MET + b_{6\,BID} BIDB + b_7 WAC + b_8 LOP + b_9 TRE + b_{10} FAR.$$

WTP_{EPWS}	:	Respondents' willingness to pay for existing public water supply,
SEX	:	Sex of the respondent (1 for male and, 0 for otherwise)
AGE	:	Age of the respondent (years)
SIZ	:	Family size (nos.)
EDN	:	Respondents educational standard (1 for literate and 0 for otherwise)
MET	:	Monthly income of the households

BID	:	Bid amount (Rs.)
WAC	:	Water colour (1 for Good colour and 0 for otherwise)
LOP	:	Low pressure (1 for low pressure and 0 for otherwise)
TRE	:	Water Treatment (whether the respondent treated water before drinking (1 for treated and 0 for otherwise)
FAR	:	Distance between residence and water source (meters)
LOC	:	Location (1 for North Chennai and 0 for otherwise)

Results and Discussion of WTP for Existing Public Water Supply

It is interesting to see that six of the eleven predictor variables regressed given negative coefficients. AGE, SEX, MET and TRE show higher negative coefficients as compared to SIZE and BID (See Table 7.27). The model revealed that increase in the value of all the six predictor variables negatively affects the WP for the existing public water supply. Many of these variables are the socio-economic attributes of the households. Only EDN has a positive coefficient which means increased EDN increase the probability of WTP. While all the indicators of unreliability taken as the prediction variables such as FAR, LOP and WAC gave positive coefficients. The possible explanation is that those households face more the problems of unreliability.

The marginal effects of the significant variables in the model show that a unit increase in AGE, SIZ, BID, SEX,MET and TRE the odds of the event (WTP) is decreased. While marginal values of FAR and LOP show that the odd of the event is significantly increased by the increase in the value of these variables. The likelihood ratio indicates that the amount of variation is different from zero. The $\div^2$ value of the model show that selected variables predicts 26 per cent of the cases willing to pay.

Table 7.27. Logit Multi-Regression Results for WTP for Existing Public Water Supply

Variable	*Coefficient*	*Marginal effects*	*T-value*	*P [|Z|>z]*
Constant	-2.809048	-.120762	-2.169	.0301
SEX	-.573774	-.246681	-.813	.4163
AGE	-.622125	-.622125	-.041	.9677
SIZ	-.127265	-.547148	-.916	.3594
EDN	.720517	.309770	1.731	.0834*
MET	-.565114	-.242958	-1.231	.2183
BID	-.129746	-.557817	-3.883	.0001***
WAC	.567768	.244099	1.806	.0709*
LOP	1.346731	.578999	2.448	.0144***
TRE	-.513662	-.220837	-.979	.3274**
FAR	.166492	.715798	4.198	.0000***
LOC	.375065	.161251	.925	.3552

Log Likelihood Value =-100.5981
Chi-Square = 73.39481
Significance level = .0000000
Psudo R^2 =0.26
No.of observation: 175

The logit model estimates the predicted probabilities of willing to pay for the existing service. If the predicted probability of predictor variables greater than 50 per cent, the household is classified into the category of WTP, otherwise it is classified as non-willing to pay household.

Model IV WTP for Improved Public Water Supply

A logit regression model was run to predict the probability of WTP for the improved public water supply.

The econometric expression of the regression equation is given as:

$$WTP_{EPWS} = b_0 + b_1 SEX + b_2 AGE + b_3 SIZ + b_4 EDN + b_5 MET + b_{6\,BID} BIDB + b_7 WAC + b_8 LOP + b_9 TRE + b_{10} FAR.$$

WTP_{EPWS}	:	Respondents' willingness to pay for existing public water supply,
SEX	:	Sex of the respondent (1 for male and, 0 for otherwise)
AGE	:	Age of the respondent (years)
SIZ	:	Family size (nos.)
EDN	:	Respondents educational standard (1 for literate and 0 for otherwise)
MET	:	Monthly Income of the households
BID	:	Bid amount (Rs.)
WAC	:	Water colour (1 for good colour and 0 for otherwise)
LOP	:	Low pressure (1 for low pressure and 0 for otherwise)
TRE	:	Water treatment (whether the respondent treated water beforedrinking (1 for treated and 0 for otherwise)
FAR	:	Distance between residence and water source (meters)
LOC	:	Location (1 for North Chennai and 0 for otherwise)

Table 7.28. Logit Multi-Regression Results for WTP for Improved Public Water Supply

Variable	*Coefficient*	*Marginal effects*	*T-value*	*P [|Z|>z]*
Constant	-2.81647	-.235330	-4.783	.0000***
SEX	-.45546	-.399696	-.304	.7610
AGE	-.24349	-.213676	-.753	.4514
SIZ	.39830	.349534	1.274	.2026
EDN	1.28391	.112675	1.383	.1666
MET	-.44519	-.390681	-.012	.9904
BID	1.07088	.939763	4.863	.0000***
WAC	3.04341	3.043414	3.937	.0001***
LOP	4.01310	4.013105	3.440	.0006***
TRE	-.50939	-.447024	-3.648	.0003***
FAR	.53381	.533812	4.532	.0000***
LOC	-3.25336	-.285501	-3.008	.0026***

Log Likelihood Value = -27.02399
Chi-Square = 494.0247
Significance level = .0000000
Psudo R^2 =0.90
No.of observation: 175

In the case of WTP for improved hypothetical water supply for public distribution points, SEX, AGE, and MET negatively influence the WTP. More household having higher household income do not want to pay for improved public water supply because they can go for in-house water connection in the immediate future. Hence, monthly household income (MET) affect negatively to the decision of willingness to pay improved public water supply. (See Table 7.28) All other predictor variables belong to the unreliability show a positive influence on the WTP. This is clearly reflected in the high marginal effects of these variables.

Having good perception about the quality of improved water (WAC) increases the probability of its usage. As expected, treating water from unimproved sources (TRE) has the tendency of decreasing the likelihood of households using improved water sources, *ceteris paribus*.

If households decide to channel resources (time and money) into treating unimproved water to render it safer, then the less likely they would use water from improved sources. Another variable, which has significant positive impact on the WTP, is the level of water pressure (LOP) to the households. The households are willing to pay more for high level of water pressure from improved public water source than the existing public water source. The other variable significantly influencing the WTP value is the distance between residence (FAR) and the public water source. FAR the distance between residence and the water fetching point, more is the probability of willing to pay for the improved water supply. The R^2 value of the model explains that the regressed variables predict 90 per cent of the cases willing to pay. Moreover, the $\div^2$ value if highly significant at zero per cent significance level.

By comparing the R^2 value of model II and IV and the highly significant $\div^2$ value at zero per cent significance level, the null hypothesis is rejected and alternative hypothesis is accepted as the research hypothesis.

Null Hypothesis (H_o):

Household WTP for new improved hypothetical water supply is not more than the current water supply.

Alternative Hypothesis (H_a):

Household WTP for new improved hypothesis water supply is more than the current supply.

The overall fitness of the logit model I and III and II and IV with 10 predictor variable revealed that the probability of WTP for new improved water supply is more than the existing water supply of households with in-house water connections and without in-house water connection by comparing the R^2 values, significant $\div^2$ values at 0 per cent significance accepted and the null hypothesis is rejected.

Null Hypothesis (H_o):

There is no significant difference in the percentage share of households without in-house water connection WTP for improved water supply as compared to households with in-house water connection.

Alternative Hypothesis (H_a):

There is a significant difference in the percentage share of households without in-house water connection WTP for improved water supply as compared to households with in-house water connection.

It becomes very clear that the percentage share of households WTP for improved water supply is more when compared to the percentage share of households WTP for the existing water supply. Moreover, households WTP for improved public water supply have also shown the same result; nevertheless the intensity is higher for the latter. The variables included models predict the WTP of public water supply much better than the in-house water supply.

8

Summary of The Findings and Suggestions

Water is an indispensable natural capital to enjoy better health. No other natural resource other than water has had such an influence on humans. When the supply of water falls short, water crisis begins. Recognizing the importance of drinking water and non-availability of substitutes at the global level, access to water became a fundamental right of people. Water scarcity has the potential to stifle developmental efforts in Chennai City. Water has always been a scarce commodity in Chennai City owing to its rapid growth of population, consecutive droughts, lack of efficient water conservation measures and poor water management. The water crisis in Chennai City is due to lack of water resources and the availability of adequate quality and quantity at the right places at the right time to meet the basic needs of the stakeholders. The water crisis involves social cost that affects the economy and quality of life.

The situation has become so appalling due to water supply restrictions, poor water pressure, more dependence on public water supplying points and unpredictability of water supply. This has led to a situation when water supply unreliability is very acute in Chennai City. At the household level, water supply unreliability is of much concern because of periodic and seasonal water scarcities experienced in the entire city. Water reliability is multifaceted, dynamic in nature and it cuts across various issues such as availability, accessibility, usage and quality. This dynamics over time and space determines the overall water security. It is the crux of the issue taken up in this study. Based on the empirical findings of the study, this chapter provides a summary of the major research findings and makes contributions to the body of knowledge in the field of economics of urban household. Recommendations to take policy measures are also proposed.

Summary of the Findings

The study comes up with the following findings:

The study reveals that 90.2 per cent of the heads in the sample households are males. The proportion of male heads in North Chennai is 89.3 per cent when compared to 91.1 per cent in South Chennai. The age group claiming the largest share is those heads aged between 40 to 59 years of age. The survey shows that there are 1945 members in 450 households with an average household size of 4.32 persons. The Backward Caste households represents nearly half (49.8 per cent) of the households. The second largest community living in Chennai City is the Scheduled Castes accounting for 28 per cent.

The overall proportion of literate heads accounts for 78.9 per cent. Illiterate heads account for 21.1 per cent comprising 25.8 and 16.4 percentages respectively in North Chennai and South Chennai. The nuclear families form the largest group accounting for 79.3 per cent. The joint families account for less than two-fifths of the households. Joint families are much predominant in North Chennai.

It is found that 29.1 per cent of the households earn an income between Rs. 2500 and Rs.5000 per month. Another 19.3 per cent households earn less than Rs.2500 per month. Thus nearly half (48.4 per cent) of the sample households earn less than Rs.5000 per month. Households with the highest monthly income of Rs.15000 and above constitute 17.1 per cent with an average income of Rs.21677.4 and a standard deviation of 7944.969. Households in South Chennai are economically much better off than the households in North Chennai. It is found that 69.8 per cent of the households own the houses occupied by them. The data indicate that three fifth (60 per cent) of the households live in independent houses; 30.9 per cent live in independent storeyed buildings and another 8.9 per cent in apartments. Out of 450 sample households, 77.3 per cent live in concrete roof houses.

The floor area of houses occupied by the sample households reveal that the average floor area of dwelling is computed as 877.11 sq.ft. It is found that 116 of the 450 households (25.7 per cent) possess showers in their bathrooms and 23 households out of the 450 (5.1 per cent) have flushing toilets in their dwellings. It is found that nearly half of the sample households (49.1 per cent) avail the facility of kitchen sink. With regard to the number of bathrooms, it is learnt that 438 of the 450 households (97.3 per cent) have bathrooms. It is usual for even the slum-dwellers to posses a bathroom covered by thatched roof. More than half (58.6 per cent) of the households have piped water connection. Nearly three-fifths (59.1 per cent) of the households in South Chennai have in-house water connection as compared to 40.9 per cent households in North Chennai.

It is interesting to note that nearly two-fifths (38.8 per cent) of the sample households are not connected to in-house water supply. Hence, they depend on public water distribution points to meet their water needs. It is noteworthy that 75.2 per cent of such households in North Chennai depend on public hand pumps as compared to 56.5 per cent in South Chennai. The disadvantage of hand pumps is that the pumps are not in good working condition or when the water pressure is low, users have to exert over a long time to get water.

Many of the households adopt alternative strategies to in-house water supply. As many as 185 sample households (41 per cent) adopt some alternative strategies to meet their water needs. Out of them 63.2 per cent adopt bore wells, 12.4 percent have dug wells and 24.3 per cent depend on packaged drinking water.

All the households connected with in-house piped connection belong to the highest monthly income slab of Rs.15000 and above. It is also revealed that the percentage of households with un-metered in-house piped water connection increases as the level of household income increases. It is also learnt that nearly two-fifth (37.9 per cent) of the households belonging to the highest monthly household income group of Rs.15000 and above rely on their own dug wells or tube-wells as alternative strategies.

The percentage of households consuming more than 60 to 80 lpcd of metro water is 39.8 in North Chennai as compared to 21.1 per cent in South Chennai. Though the socio-economic and demographic fundamentals of households in South Chennai are strong when compared to the North Chennai, the average lpcd of metro water consumed by the households in South Chennai is less in comparison with the lpcd of North Chennai. It may also be due to the fact that the practice of marketisation of water in North Chennai appears less while marketisation of water is very strong in South Chennai. One of the important inferences of the study is that the dependency on public water supply is more among the poor households. Illiterate heads form around one-fourth of the total households, who consume more than 120 lpcd of metro water. However, two-fifths (40 per cent) of the households in the highest monthly income slab of Rs.15, 000 and more avail public water to the levels of 20 to 40 lpcd. Contrary to this, more than half of the households belonging to the lowest monthly household income level of below Rs.2500 consume public water to the tune of 60 – 80 lpcd. Household size records a positive association with lpcd of metro water. For instance, the maximum number (38.5 per cent) of households in the family size of 6-8 represents the water consumption category of 80-100 lpcd. It is interesting to see that households using bath showers are in the category of consuming less lpcd of public water, i.e., 80 per cent of such households consume less than 40 lpcd. Similarly households having more flushing toilets use less lpcd of metro water when compared to the households having single flushing toilets. More than two fifths (44.6 per cent) of the households with single flushing toilet consume 60-80 lpcd; 29.8 per cent of the households having two flushing toilets consume 60-80 lpcd.

A maximum number of households using one and two kitchen sinks consume 60-80 lpcd, and they are in the percentages of 32.4 and 29.5 respectively of the total households using kitchen sinks. It is revealed that a majority of the households have single washbasin and the number of households decrease when the number of washbasins per household increase. The number of bathrooms per household varies from one to four in the percentages of 68.5, 27.9 and 0.7 respectively of the 93 per cent households with bathrooms. A maximum number of households with single bathroom (45.3 per cent) consume 60-80 lpcd of water. It is found that more than two-fifth of the households (41.5 per cent) consume 60-80 lpcd of water, of which 43 per cent live in the houses having an area of 300 to 600 sq.ft. Hence, it becomes quite clear that the households in the lower socio-economic ladder depend on metro water to meet their domestic requirements while the richer households find some alternative strategies besides purchasing water from the market.

In order to examine the differences in the water consumption pattern of households with respect to metro water and private water, the per capita per day total water consumption of the sample households are compared with various social and economic factors. The lpcd of total water consumption of the sample households in Chennai City shows that nearly half (47.1 per cent) of them use 50 to 75 lpcd. It is found that 80.5 per cent of the households with a family size of 4 to 6 members consume water between 50 lpcd and 125 lpcd. It is quite clear that the maximum number of households in all the income classes except the highest income class is found in the lpcd level of 50-75 litres. Hence it is evident that there exists a positive relationship between total water consumption and monthly household incomes. But this relationship could not be established in the case of metro water.

The determinant of the per capita metro water consumption is regressed by taking six independent variables. The value of adjusted R^2 is 0.220, indicating that 22 per cent of the variance of the per capita metro water consumption per day is explained collectively by the six variables. The results indicate that a 1 per cent increase in per capita metro water consumption per day is associated with 7 per cent decrease in kitchen sinks, 5 per cent increase in the household size, 2 per cent decrease in the number of showers and 1 per cent decrease in the number of flushing toilets. The Durbin Watson statistic is estimated as 1.759(d). Since d is less than dL the null hypothesis is rejected. Hence the selected variables explain the determination of per capita metro water consumption per day by the household. Since 4 – 1.759 e" du, it can be concluded that (d = 2.241) there is no negative autocorrelation. Since the computed 'F' values of all the six predictor variables are higher than the values given in the F-distribution table at 1 per cent level of significance, the null hypothesis is rejected and the accepted variables used in the model are known to assert and influence the per capita metro water consumption per day of the selected households in Chennai City.

Survey reveals that only 1.5 per cent of the households with in-house water connection have water for 12 hours or more each day. Such households are located very near to the main distribution points. More than a quarter (28.8 per cent) of the households with in-house water connections has less than 2 hours of water per day. Nearly half of the households (46.2 per cent) receive 2 to 4 hours of water per day. The duration of water supplied every day has been very low. Therefore, the average duration of supply in summer and winter does not make any big difference. The households with in-house water connection in North Chennai receive lower average duration of Metro water supply compared to the same type of households in South Chennai. The Pearson's correlation between monthly household income and the duration of water supply per day of the households with in-house water connection has been computed as 0.155. The value of correlation coefficient shows a relatively low degree of positive correlation between these two variables. It is revealed that the proportion of households with in-house water connection getting water in good pressure is a little over 15 per cent. A majority of the households (47.3 per cent) receive water at an average pressure followed by those with poor pressure, accounting for 35.6 per cent.

The inequalities due to pressure are more severe in North Chennai because good pressure accounts for only 0.9 per cent of households. The percentage of households with poor pressure is 56.6 and those with average pressure exceeds two-fifths (41.7 per cent) of the households with in-house connections. The duration of water availability for the households with in-house water connection is short characterized by low and irregular water pressure. It is inferred that more than half (54.2 per cent) of the households fetch Metro water in the morning hours. Almost one-fourth of the households (24.2 per cent) with in-house water connection receive water in the night. Both morning and night are preferred because such time is convenient for the households to make the necessary arrangements without disturbing their regular work. But on the other hand, this would give a lot of hardship to them during the office time due to lateness of the night.

Households with in-house water connection also face severe unreliability problems like short duration of supply, low pressure and un-predictability that led them to invest in alternative strategies. Often they purchase water as a temporary measure. Water pressure in the distribution points is one of the important determinants of unreliability for the households without in-house water connection. North Chennai has 64.6 per cent of the total households without in-house water connections. The quality of water pressure in North Chennai is comparatively poor. The proportion of households receive water in good pressure is only 12 per cent, which is almost the same for both North Chennai and South Chennai. On an average nearly half (48.6 per cent) of the households without in-house water connection have average water pressure. The water pressure indicator reveals that 42.5 per cent of the households in North Chennai face poor pressure as compared to 25.8 per cent of households in South Chennai. No household in South Chennai represents in the category of *'no pressure'*.

The proportion of households served with continuous water supply all the days in a week accounts for 2.3 per cent of the households without in-house water connection. However, three-fifths (57.7 per cent) of the households are supplied with water for four days in a week. In North Chennai, 62.8 per cent of households are supplied with water four days in a week. More than four fifths (79.4 per cent) of the households without in-house water connection belong to the monthly income category of less than Rs.5000. More than (57.7 per cent) of the households without in-house water connection avail water supply four days in a week. A majority of them represents the lowest monthly income categories of below Rs.2500 and Rs.2500-5000.

Intermittent water supply is termed as one of the important indicators of unreliable water supply. The interruption in water supply is frequent for the households, which fetch water from the street sources. Unpredictability of water availability is extreme in South Chennai because 29 per cent of the households getting water from common sources do not know the time of water availability. The mean time of travel for water collection by the households per day is estimated at 0.2503 hours, with a standard deviation of 0.11289. The mean time spent by households for waiting per day is computed as 2.2691 hours with a standard deviation of .69088. Households without in-house water connection spend considerable time every day for water collection.

More than one-quarter (25.7 per cent) of the households without in-house water connections spend less than 10 minutes per head per day for water collection followed by 23.43 per cent spend 20 minutes, 25.71 per cent spend 30 minutes and 25.14 per cent spend 40 minutes. Therefore, the number of household decrease when the time for water collection increase. It is generally found the travel time for collecting water changed according to the changes in the income of the households.

The in-depth study of water supply unreliability reveals that the poor income households without in-house connection face extreme level of unreliability problem due to unpredictability of water supply, low level of supply, long duration of time spent in collecting water and their inability to adopt compensatory strategies, as these are expensive.

It is also revealed that three-fifths (60 per cent) of the households adopt one or more compensatory strategies. More than three-fourths of the households have (75.1 per cent) stored water in *Kodams* (plastic pot) and buckets, which are widely prevalent compensatory strategies. Mostly poor households use such a practice as a measure to reduce the problems of unpredictability of water supply. As many as 27.3 per cent households have bore wells and they have built complementary overhead tanks in order to store water, especially for bathing and washing purposes. Due to the low pressure at the tap, a majority of households (47.7 per cent) installed hand pumps to draw water. It is reported that 4 per cent of the respondents shift their residence due to shortage of water. Over 10 per cent of the households adopt compensatory strategies such as construction of underground tank to store Metro water and then lift it to the overhead tanks. It is interesting to note that 60 per cent of the households treat water before cooking and drinking. The households in North Chennai (85.3 per cent) depend more on the strategy of storing water in kodams when compared to the households in South Chennai (64.8 per cent).

Households use underground and overhead tanks to distribute purchased private water, metro water, and water from dug wells and bore wells. Some well-to-do households have separate underground and overhead tanks for metro water and shallow water. It is revealed that only 10.9 per cent (49 out of 450) of the households have underground and overhead tanks to store Metro water. Nearly two thirds (65.3 per cent) of the households possessing underground water sumps are in the highest monthly income bracket of Rs.15000 and above. Many areas of Chennai City experience insufficient water pressure in in-house water connection.. Hence, the underground water sumps constructed for water storage remains normally empty. It is found that 48.4 per cent of the households having in-house hand pumps are earning a monthly income of less than Rs.5000 in South Chennai. But in North Chennai more percentage of households is found in this income group.

Dug well strategy is one of the oldest strategies to supplement the water needs for bathing and washing. Out of the 450 households, only 2.5 per cent have their own dug wells, however bore wells are the common strategy used by the households in Chennai City to tide over the unreliable water supply. It is found that 27.3 per cent of the households have installed bore wells. Over three fifths (64.2 per cent) of the bore wells are in South Chennai. It is inferred that 39.8 per cent of the households possessing bore wells and the associated overhead tanks represent the highest monthly income slab of Rs.15, 000 and more.

It is reported that 60 per cent of the sample households treat water for drinking purpose everyday. But this practice is done more by the households in South Chennai (67.1 per cent) as compared to the 52.9 per cent of the households in North Chennai. The average cost of unreliability per household is estimated as Rs.7838.93 per year. But the households with in-house water connection pay Rs.600 per year as water tariff i.e. 13 times less than the estimated average *'unreliability cost'*. It is evident that 47.6 per cent of the households have a family size ranging from 2-4 members, of whom nearly a half (47.2 per cent) incurs an average annual unreliability cost of less than Rs.5000. But, when the family size increases, relatively less number of households is found in the average unreliability classes of less than Rs.5000. The Karl Pearson's correlation coefficient between the averages of unreliability cost per household and the size of the household is estimated as 0.138 at 1 per cent significance level, which exhibit a relatively low degree positive correlation.

The highest monthly household income class of Rs.15, 000 and above constitutes 77.5 per cent of the households in the unreliability cost classes of less than Rs.7500. While 91 per cent of the households in the highest average unreliability cost of Rs.17500 and above is represented in the income group of less than Rs.7500 per month. The Karl Pearson's Correlation coefficient value of -0.33 at 0 per cent significance level shows that the average unreliability cost decreases with every increase in the monthly income of households. It was hypothesised that there is no association between literacy levels of the respondents and the average unreliability cost.

The $\div^2$ test verified the existence of a significant difference between the metro water consumption of households with in-house water connection and without in-house water connection. The OLS multiple regression model did not accept household income as one of the determinants of lpcd consumption of metro water. Hence, the null hypothesis is accepted that the household income has no influence on the metro water demand. The Karl Pearson correlation coefficients register negative value between household income and area of house with average annual household unreliability cost. There is a negative association between household income and inefficient alternative strategies of storing water. It means that poor households adopt inefficient alternative strategies when compared to rich households.

The study makes it clear that a majority of the sample households is not happy with the existing water supply. This is the reason why more people are wiling to pay an amount higher than the existing tariff for the hypothetical package of water supply. The urban poor depend more on metro water hence they face a higher unreliability cost when compared to the households depend on the private water market. There should be some positive actions aimed at looking the water needs of the poor. Availability of safe drinking water is not the business of the individuals but it is the responsibility of the government. These factors are being stressed in order to emphasize the need for reliability and accessibility to all the households all the time. The reliability and accessibility is nothing but the sustainability in making available water to the largest number and not just the affluent few their luxurious lifestyles.

Suggestion

A better policy decision to provide reliable water supply and its universal accessibility is determined by a clear understanding of the behaviour of different types of households as consumers. The following are the suggestions to ensure reliability and universal coverage of residential water supply though an effective demand side management which can mitigate the water crisis in Chennai City.

1. Safe water is indispensable for the healthy and productive life of all the urban dwellers. Adequate water supply reduces public health risks that ease the fiscal responsibility of the state on medical expenditure. Metro water board should take special care in ensuring reliable and adequate water supply to slum households by investing sufficient amount in infrastructural facilities in order to eliminate the system of rationing of water.
2. Urban poor, especially the slum dwellers bear the unreliability cost to the maximum. They depend on metro water for the essential as well as their supplementary needs. Hence alternative water provisions should be made for the supplementary water needs viz., bathing, cleaning etc. The supplementary water provisions include the digging of bore wells, community wells etc., for every 50 slum households. It will considerably reduce the unreliability cost and the wastage of precious safe water.
3. The current water pricing policy of Chennai City is one of the administered pricing or flat rate tariff which is uneconomic and prone to the misuse or over exploitation misuse of water resources. Flat water tariff helped the well–to–do households the maximum when compared to the poor. Even the poor households express their willingness to pay for reliable water. Hence a progressive water pricing policy should be devised for a reliable water supply to the residential sector by taking volume of water consumption and affordability as the two important determinants of water pricing.

4. Rationing or mobile supply of water is uneconomic or temporary measure to tide over the water crisis. However, it should be gradually substituted by pipe ways. This requires a master plan by assessing the existing piped network, additional requirements and the necessary replacements on a long-term basis. Metro Water Board should have a plan of finance and recovery in the process of universal coverage of water supply in Chennai City.
5. Efforts should be made for the dissemination of better water use and hygienic practices, community education and awareness building. These are the cheapest and the most efficient components of water demand management.
6. Metro Water Board should have a mechanism to control the private sector in the water market. It should test and monitor the water supply standard and encourage competition in the water market.
7. The metro board should have a water distribution station in each corporation division in order to ensure reliable water supply to receive complaints and to rectify the complaints within a shortest span of time.
8. Metro Water Board should supply packaged water at a fair price in order to compete with the private players in the water market.
9. Water planning, water audit, and assessing the performance of various divisions should become a periodical affair.
10. The procedure and formalities to get new in-house water connection should be simplified.
11. Efforts should be made to control the loss and misuse of metro water.
12. Repair and maintenance of public water stand posts should be on an emergency basis.
13. Metered in-house water connections should be encouraged.
14. Metro water planning should incorporate alternative water sources.

15. Chennai city is already declared as the most severe water starving metropolitan city in India. Hence reliable water supply will remain as a dream if there is no efficient supply side water management. In order to improve the water supply the following measures need to be undertaken.
    (a) Increase the storage capacity of water reservoirs in order to accommodate the entire rain water during the monsoon season.
    (b) Methods are to be followed for the prevention of evaporation of surface water.
    (c) Rain water harvesting should be made rigorously. All public buildings and roads should be brought under rain water harvesting methods.
    (d) Water recycling and reuse in all the industrial establishments should be encouraged.
    (e) Sewage water can be treated for the industrial uses.
    (f) Urban planners should take measures to preserve aquifers, wetlands agricultural and open spaces in and around Chennai City which are essential for the survival of the city.
    (g) Withdrawal of underground water should be quantified and limits should be prescribed on pumping in order to maintain sustainability of water supply.
    (h) Measures should be taken to avoid water pollution and mixing of metro water with sewerage pipes.

Bibliography

BOOKS

[1] Archibald, S.O. and M.E. Renwick (1998): "Expected Transaction Costs and Incentives for Water Market Development", Markets for Water: Potential and Performance, ed. K.W. Easter et al., Boston MA: Kluwer Academic Publishers.

[2] Ashima, Goyal (2004): "Water" Getting water from public-private partnerships India Infrastructure Report", Oxford University Press New York.

[3] Aswathanarayana.V. (2001): "Water Resource Management and the Environment", A. A. Balkema Publishers, Japan.

[4] Chattopadhyay Subrata (2001): "A Practical Approach to Water Supply and Sanitation in Asian Cities", Published by CombiGrafik, Sweden.

[5] Chetan, Vaidya (1998): "Willingness to pay for water and Sanitation Services: A case study of Baroda", HSMI Publication, New Delhi.

[6] Chris Park, (1997): "The Environment" Principles and Applications, Routledge Publication, London.

[7] Coing H, de Lara P and Montano I (1989): "Privatisation et R®™ulation des Services Urbains: U\une etude Comparative (Argentine, Brø™il, Sí™®™al), Plan Urbain", Paris: LATTS, ENPC.

[8] Cosgrove, Williams and Rijisberman, Frank. R, (2000): "World Water Vision: Making water everybody's Business", Earth scan Publication, London.

[9] Deaton. A, (1997): "The Analysis of Household Surveys: A Micro-econometric Approach to Development Policy", Baltimore, Maryland, The Johns Hopkins University Press.

[10] Felix Clodds, (2002): "Fresh Water: A Global Crisis of Water Scarcity and Basic Water Provisions", Earth Scan publication Limited, London.

[11] Geoffrey.D. Datoelko (2002): "Finding the source, the linkages between population and water", Woodrow Wilson International Center for Scholars, Environmental Change and Security Project, CSP, Publication University of Michigan, USA.

[12] Gleick, P. -H. 1993. 'Water in Crisis: A Guide to the World's Fresh Water Resources', Oxford University Press, New York.

[13] Hanemann, M.W., (1998): "Determinants of urban water use", in Bauman, d., Boland, J., and Hanemann, M.W. (eds.) Urban Water Demand Management and Planning, McGraw-Hill Publication.

[14] (1998a): "Determinants of Urban Water Use", Urban Water Demand Management and Planning, ed. F. Baumann, J. Boland, and W.M. Hanemann, New York: McGraw-Hill.

[15] Hanley, N., J. F. Shogren and B. White (1997): "Environmental Economics in Theory and Practice", Oxford University Press, New York.

[16] Hardoy, J.E., D. Mitlin and D. Satterthwaite (1992): "Environmental Problems in Third World Cities", London: Earth scan Publications Ltd.

[17] (1990): "The poor die young: housing and health in third world cities", London, Earth scan publication.

[18] Jacob Johno (1997): "Economics of Public Water Supply System with special reference to Kerala", Rawat Publication, Jaipur and New Delhi.

[19] John De Zugne, P.E. (1996): "Hand Book of Drinking Water Quality", published by Vgn Nostrand Reinhold Publication, New York.

[20] Kumaraswamy. T. (1996): "An Integrated Approach to the Drinking Water Problem of Chennai City", India, Environmental Research Forum Vols, 3-3(1996), Transtec Publications, Switzerland.

[21] Kundu, Amitabh (1992): "Urban Development and Urban Research in India", Published by Anupam Banerjee for Khama Publication, New Delhi.

[22] Mathur. R. John, (1984): "Water Resource, Distribution, Use and Management", V.H. Winston and Sons publication, New York, USA.

[23] Maude Barlow and Tony Clarke (2002): "Blue Gold, The fight to stop the corporate theft of the world's water", Published by Left World, New Delhi.

[24] McFadden, D., (1974): "Conditional Logit Analysis of Qualitative Choice Behavior", in P. Zarembka, ed., Frontiers in Econometrics, Academic Press, New York.

[25] Mehta M (1999): "A Review of Public-Private partnerships in Water and Environmental Sanitation in India", New Delhi: Department for International Development. Prevention and Food Adulteration Act (1954) 2nd Edition, New Delhi: International Law Book Company.

[26] Montgomery, Mark, R. Stren, B. Cohen, and H. Reed (eds.) (2003): "Cities Transformed: Demographic Change and Its Implications in the Developing World", National Academy Press, Washington, D.C.

[27] Munasinghe, Mohan. (1979): "The Economics of Power Systems Reliability and Planning", Published for the World Bank, The Johns Hopkins University Press, Baltimore.

[28] Nageswara Rao, (1993): "Economics of Water Supply, A case of water supply undertakings of the urban centers of Rayalasema", Himalayas Publishing House, New Delhi.

[29] Nickum, J.E., and K.W. Easter, (1994): "Metropolitan Water Use Conflicts in Asia and the Pacific", West View Press, Boulder, CO.

[30] Okun. D. (1998): "The price of water in Urban areas of Developing Countires' in Couts et prix del' eau en ville", Aetes de colloque. Presses de l' Ecoloe Nationale des Ponts et Chaussess, Paris.

[31] Postel, S. (1993): "Water in Crisis: A Guide to the World's Fresh Water Resources", New York, Oxford University Press.

[32] Prabhkar V.R. (2001): "Environment impact Assessment: "Water Environment", Anmol Publications Pvt. Ltd., New Delhi, India.

[33] Ratna Reddy (1996): "Urban Water Crisis", Rationale for Pricing, published by Prem Ratwat publication, Jaipur, India.

[34] Ravindra Prasad.D and Malla Reddy A, (1994): "Environmental Improvement of Urban slums: The Indian Experience", Regional Center for Urban Environmental Studies Osmania university, Book links Corporation Publication, Hyderabad, India.

[35] Robert B. Dean and Ebba Lund (1981): "Water Reuse: Problems and Solutions", Published by Academic Press, London.

[36] Swades Kumar Bhattacharya, (2003): "Urban Domestic Water Supply in Developing Countries", CBS Publishers and distributors, Delhi.

[37] Swapam C. Dob, (2003): "Environment Management", Jai co Publishing House, Chennai, India.

[38] UN-Settlement Programme, (2003): "Water and Sanitation in the World Cities", Local action for global goals, Earth scan publication, London.

[39] Water Lorch, (1982): "Handbook of Water Purification", McGraw-Hill Book Company (UK) limited, London.

[40] White G F, Bradley D J, White A. U, (1972): "Drawers of water: domestic water use in East Africa", University of Chicago Press, Chicago.

[41] Whittington, Dale, et al., (1991): "Contingent Valuation: Estimating the Willingness to Pay for Housing Services: A Case Study of Water Supply in Southern Haiti", Tipple, A. Graham, Willis, Kenneth G. (eds.) Housing The Poor In The Developing World: Methods of Analysis, Case Studies And Policy. London and New York: Routledge.

[42] WHO/UNICEF, (2000): "Fresh Water: A Global Crisis of Water Scarcity and Basic Water Provisions", Earth Scan Publication Limited, London.

[43] Zû™ah, Marie-HØ™)'e, (2000): "Water: Unreliable Supply in Delhi", Manohar Publishers, New Delhi.

[44] Khandwalla, Pradip (1999), Revitalizing the State: a Menu of Options (New Delhi: Sage, 1999a).

JOURNALS

[1] Agthe, D. and B. Billings (1980): "Dynamic Models of Residential Water Demands", Water Resources Research 16, No. 3.

[2] Alberini, A. (1995): "Efficiency Vs Bias of Willingness-to-estimates Bivariate and Internal-Data Models", Journal of Environmental Economics and Management, 29.

[3] (1997): "Modeling Responsive Incentive effects in Dichotomous Choice Contingent Valuation Data", Land Economics, 73.

[4] Altaf Mir Anjum, (1994): "The Economics of Household Response to Inadequate Water Supplies", Evidence from Pakistan', Town and WPR, Vol. 16, No. 1.

[5] Altaf, A., Jamal, H. and Whittington, D. (1992): "Willingness to Pay for Water in rural Punjab, Pakistan", UNDP-World Bank Water and Sanitation Program, World

Bank, Washington, D.C. (1818 H St., NW, Washington 20433).

[6] An Y.M, (2000): "A semiparametric distribution for willingness to pay and statistical inferences with dichotomous choice contingent valuation data", Journal of American Agricultural Economics Association. Vol. 82.

[7] Armstrong, R.D. and Willis, C.E. (1977): "Simultaneous Investment and Allocation Decision Applied to Water Planning", Management Science, Vol.23, No.10.

[8] Butcher W.S. Haunes, and Hall W.A., (1969): "Dynamic programming for the optimal sequencing of water supply projects", Water Resources research, Vol.5, No.6.

[9] Cameron, T.A. and D.D. Huppert (1988): "OLS versus ML Estimation of Non-market Resource Values with Payment Card Interval Data", Journal of Environmental Economics and Management, Vol. 17.

[10] (1987): "Efficient Methods for Closed-ended Contingent Valuation Surveys", The Review of Economics and Statistics, Vol. 69.

[11] Creel, M. (1998): "A note on consistent estimation of mean WTP using a misspecified logit contingent valuation model", Journal of Environmental Economics and Management 35(3).

[12] Creel, Michael, (1998): "A Note on Consistent Estimation of Mean WTP Using a Misspecified Logit Contingent Valuation Model", Journal of Environmental Economics and Management; 35(3), May.

[13] Dandy, G., T. Nguyen, and C. Davies (1997): "Estimating Residential Water Demand in the Presence of Free Allowances", Land Economics 73, No.1.

[14] Darr, P., S.L. Feldman, and C.S.Kamen (1975): "Socioeconomic Factors Affecting Domestic water Demand in Israel", Water Resources Research 11 (6).

[15] David L. Chicoine, Ganapathi Ramamurthy, (1986): "Evidence on the Specification of Price in the Study of Domestic Water Demand", Land Economics, Vol. 62, No.1.

[16] Gibbs, K.C., (1978): "Price Variables in Residential Water Demand Models", Water Resources Research, 14(1).

[17] Goett, A., K. Hudson, and K. Train, (2000): "Customers' Choice among Retail Energy Suppliers: The Willingness-to-Pay for Service Attributes", The Energy Journal, Vol. 21, No. 4.

[18] Gottlieb, M., (1963): "Urban Domestic Demand for Water: A Kansas Case Study", Land Economics, 39(2).

[19] Green, D., et al. (1998): "Referendum Contingent Valuation, Anchoring, and Willingness to Pay for Public Goods", Resource and Energy Economics, 20(2).

[20] Green, W. F. (1995): "LIMDEP version 7: Users Manual", Econometric Software Inc., Bellport, New York.

[21] Gupta, I. (2002): "Rural Migrants Segmentation and Labour: Micro-Level Evidence from Delhi Slums", Economic and Political Weekly, January, 12.

[22] Halvorsen, Bente; Saelensminde, Kjartan, (1998): "Differences between Willingness to Pay Estimates from Open Ended and Discrete Choice Contingent Valuation Methods: The Effects of Heteroscedasticity", Land Economics; 74(2), May.

[23] Hausman, J., and D. McFadden (1984): "Specification Tests for the Multinomial Logit Model", Econometrica, 52 (5).

[24] Headley, J. C. (1963): "The Relation of Family Income and Use of Water for Residential and Commercial Purposes in the San Francisco-Oakland Metropolitan Area", Land Economics. 39(4).

[25] Henry S. Foster, Jr., Bruce R. Beattie, (1981): "Urban Residential Demand for Water in the United States: Reply", Land Economics, Vol. 57, No.2.

[26] H?lund, L. (1999): "Household demand for water in Sweden with Implications for a Potential Tax on Water Use", Water Resources Research, 35(12).

[27] Howe, C.W. and Linaweaver, F.P. (1967): "The Impact of Price on Residential Water Demand and its Relationship

to System Design and Price Structure", Water Resources Research, 3 (1).

[28] Howe, Charles W. et al., (1994): "The Value of Water Supply Reliability in Urban Water Systems", Journal of Environmental Economics and Management, 26(1), January.

[29] Jaysasundara (J.), Kotagama (H.B.) and Weerahewa (J.), (1999): "Willingness to Pay for Drinking Water", Tropical Agricultural Research, (11).

[30] Jones, C.V., and J.R. Morris, (1984): "Instrumental Price Estimates and Residential Water Demand", Water Resources Research, 20(2).

[31] Jordan J.L. and A.H.Elangheeb (1993): "Willingness to Pay for Drinking Water Supply", Water Resources Research, Vol. 27.

[32] Jung, T.Y. (1993): "Ordered Logit Model for Residential Electricity Demand in Korea", Energy Economics, Vol. 15.

[33] Langford, I. H., et al. (1998): "Improved Estimation of Willingness to Pay in Dichotomous Choice Contingent Valuation Studies", Land Economics 74(1) 65-75.

[34] Larson, B.O. and H.E. Hudson, Jr. (1951): "Residential Water Use and Family Income", Journal of the American Water Works Association. 43(7).

[35] Llorenthe, Marie and Marie Helene Zerah (2003): "The Urban Water Sector- Formal Versus the Informal Water Suppliers in India", Urban India,, Volume XXII No.1, National Institute of Urban Affairs, January –June.

[36] Madanat, S., and F. Humplick (1993): "A Model of Household Choice of Water Supply Systems in Developing Countries", Water Resources Research, 29 (5).

[37] Malcolm, Farley (1999): "Managing Water Loss in Developing Countries", Water 21, July-August.

[38] March, J. G. Olsen, J. P., (1972): "A Garbage Can Model of Organizational Choice", in Administrative Science Quarterly, 17.

[39] Marie Llorente and Marie Helene Zerah Urban India, (2003): "The Urban Water Sector: Formal Versus Informal Suppliers in India", National Institute of Urban Affairs, January-June, Vol. XXII, No 1.

[40] Mart$Sez-Espi?ira, R., (2002): "Residential water demand in the Northwest of Spain", Environmental and Resource Economics, 21(2), 161:187.

[41] McPhail, Alexander A., (1994): "Why Don't Households Connect to the Piped Water System? Observations from Tunis, Tunisia", Land Economics, Vol. No.70(2), May.

[42] Pham Khanh Nam andTran Vo Hung Son (2005): "Household Demand for Improved Water Services in Ho Chi Minh City: A Comparison of Contingent Valuation and Choice Modeling Estimates", Environmental Economics Unit, Department of Economics – HCMC, Ho Chi Minh City, Vietnam,

[43] Postal, Sundra, (1985): Conserving Water, "The Untapped Alternative", World Water Paper 67, September.

[44] Quiggin, J. (1998): "Individual and household willingness to pay for public goods", American Journal of Agricultural Economics 80(1).

[45] Riordon, Courtney (1973): "Multi-state Marginal Cost Model of Investment-pricing Decision", Application to Urban water Supply Treatment Facilities, Water Resources Research, Vol. 9, No.2.

[46] Sarma, K.S.R.N. (1991): "Urban Water Supply Tariffs: Issues in Determination and Administration", CUS, Indian Institute of Public Administration, New Delhi.

[47] Sengupta, C (1999): "Dynamics of Community Environmental Management in Howrah Slums", Economic and Political Weekly, May 22.

[48] Shiklomanov, I.A. (2000): "Appraisal and Assessment of World Water Resources", Water International.

[49] Sundar, R. and A. Sharma (2002): "Morbidity and Utilisation of Healthcare Services: A Survey of Urban Poor

in Delhi and Chennai", Economic and Political Weekly, November 23.

[50] Turnovsky, S.J. (1969): "The Demand for Water: Some Empirical Evidence of Consumers' Response to a Commodity Uncertain in Supply", Water Resources Research. 5(2).

[51] UNEP Industry and Environment (2002): January-March, ISSN 0378-9993, E Volume.25&I. No.1

[52] Viswanathan.S. (2004): "Water Scarcity in Chennai", Industrial Economist, 30 August-14 September.

[53] Walced .K, Al-Zubari, (2002): "Alternative Water Resource Management Policies in West Asia", UPEP Industry and Environment, ISSN0.378-9993, Volume 25, No.1, January-March.

[54] Whittington K. and Lauria D. (1990): "Estimating the Willingness to Pay for Water Services in Developing Countries", A Case Study of the Contingent Valuation Method in Southern Haiti, Economic Development and Cultural Changes 38, No. 2.

[55] Whittington, D., Briscoe, J., Mu, X., Barron, W., (1990): "Estimating Willingness to Pay for Water Services in Developing Countries: A Case Study of the Use of Contingent Valuation Survey in Southern Haiti", Economic Development and Cultural Change.

[56] Whittington, Dale, Donald T. Lauria and Xinming Mu (1991): "A Study of Water Vending and Willingness to Pay for Water in Onitsha, Nigeria", World Development, Vol. 19 (2/3).

[57] Wong, S.T., (1972): "A Model of Municipal Water Demand: A Case Study of Northeastern Illinois", Land Econcmics, 48(1).

[58] Cairncross, Sandy (2003). Water Supply and sanitation: some misconceptions.' Tropical Medicine and International Health 8: 193 – 195.

PAPERS

[1] Alebel Bayrau (2002): "Affordability and Willingness to Pay for Improved Water Supply in Urban Areas of Ethiopia", A paper presented to the Ethiopia Federal Ministry of Water Resources and the African Economic Research.

[2] David McKenzie Isha Ray (2004): "Household Water Delivery Options in Urban and Rural India", Paper prepared for The 5th Stanford Conference on Indian Economic Development, June 3-5.

[3] Dinar A and Saleth R M (1997): "Satisfying Urban Thirst: Water Supply Augmentation and Pricing Policy in Hyderabad City", India, Technical Paper No. 395, Washington D.C.: World Bank.

[4] Harris and Clive (2003): "Private Participation in Infrastructure in Developing Countries: Trends, Impacts and Policy Lessons", World Bank Private Sector Advisory Services Department Working Paper No. 5.

[5] Humplick, F. Kudat A. and Madanat.S. (1992): "Household Response to Reliability of Water supply, The case of Istanbul, Turkey", Policy Planning and research Staff, World Bank, Washington, D.C.

[6] Humplick, Frannie. (1991): "Reliability of Infrastructure Systems: a Holistic Representation", Draft paper, Infrastructure and Urban Development Department, World Bank, Washington, D.C.

[7] Janakarajan S and Marcus Moench (2002): "Are Wells a Potential Threat to Farmers' Well Being? The Case of Deteriorating Ground Water Irrigation in Tamil Nadu", Working paper No.174 of Madras Institute of Development Studies.

[8] Kornai T. (1982): "The Shortage Economy", Amsterdam, Elsevier.

[9] Malla P, (1983): "Logit Analsyis of Technology Adoption by Rice Farmers in Dhanusha District", Nepal, Research Paper Series npÿ 22, Kathmandu, Agricultural Projects Services Centre.

[10] Munian. A (2004): "Infrastructure: Water Unreliability in Chennai City", Paper presented in the National-University Grants Commission Seminar organized by the Head of the Department of Economics, Pachiappa's Arts College for Men, Kanchipuram, Tamil Nadu, India, on 11.09.2004.

[11] Murty, M.N., S.C. Gulati and A. Banerjee (2003): "Hedonic Property Prices and Valuation of Benefits from Reducing Urban Air Pollution in India", Discussion paper No. 61/2003, Institute of Economic Growth, Delhi.

[12] Roy, J. (2004): "Willingness to Pay Studies: A Policy Tool. Paper Presented at the Conference on the Market of Water and Waste Technologies Through Environmental Economics", 28th and 29th May, Paris.

[13] Ruet, Joel Saravanan (2002): "The Water and Sanitation Scenario in Indian Metropolitan Cities: Resources and management in Delhi, Kolkatta, Chennai and Mumbai", CSH occasional Paper No.6, French Institute, Delhi.

[14] Ruet, Joel Saravanan and Zerah, Marie-Helene (2002): "The Water and Sanitation Scenario in Indian Metropolitan Cities: Resources and Management in Delhi, Kolkatta, Chennai and Mumbai", CSH occasional Paper No.6, French Institute, Delhi.

[15] Vaidyanathan. A and Saravanan. J (1997): "A Background paper from the center for science and environment", Managing Water in Chennai, Published by Center for Science and Environment, New Delhi, India.

REPORTS

[1] Altaf, Anjurn., Haroon Jamal, Dale Whittington (1992): "Willingness to Pay for Water in Rural Punjab, Pakistan", Water and Sanitation Report 4, UNDP-World Bank Water and Sanitation Program, International Bank for Reconstruction and Development, Washington, D.C.

[2] Anil Agarwal, (1997): "Dying Wisdom", State of India's Environment-A Citizens Report, Publication, Center for Science and Environment, New Delhi.

[3] Annual Report (1994-95): CMWSSB.

[4] (1995-1996): CMWSSB

[5] (1998-1999): CMWSSB.

[6] (2000 – 2001): CMWSSB.

[7] (2001-2002): CMWSSB.

[8] (2002-2003): CMWSSB.

[9] (2003-08): CMWSSB.

[10] Asian Development Bank (ADB), (1993): "Theme Paper-Service Levels and the Urban Poor", Managing Water Resources to Meet Mega city Needs, Proceedings of Regional Consultations, Manila.

[11] (1993): "Water Utilities Data", Asia and Pacific Region.

[12] Carson, R.T., and R. Mitchell. (1987): "Economic Value of Reliable Water Supplies for Residential Water Users in State Water Project Service Area. Report by QED Research", Inc., Palo Alto, CA, for the Metropolitan Water District of Southern California, Los Angeles, CA.

[13] Census of India, (2001): "Census of India (Provisional) Tables; Cities with more than one Million Population; and Urban Agglomerations/Cities having Population of one Million in 2001", Office of the Registrar General: New Delhi, India.

[14] Central Pollution Control Board (CPCB),(2002): "Water Quality in India, Status and Trends, Central Pollution Control Board", New Delhi.

[15] (2001): "Environmental Improvement of Water ways in Chennai City", New Delhi.

[16] (1991): "policy imperatives and agenda for action": (based on a research programme conducted by The Times Research Foundation) October.

[17] CMWSSB (1991): "Chennai Metropolitan Water Supply and Sewerage Board's Water Tax and Sewerage Tax (Levy Collection) Regulations".

[18] (2001): Environmental Improvement of Water ways in Chennai City, Chennai.

[19] District Environment Profile, (1996): "A study Report of Volume of Chennai District" The Times Research Foundation, October 1991.

[20] Economic Survey of India (2003-2004): "Water resources research grant proposal" Dynamic Simulation of Water Distribution Systems with Instantaneous Demands, 2000.

[21] Global Report on Human Settlements (2001), "Cities in a Globalising World, Centre for human Settlements (HABITAT), published in UK and USA.

[22] Government of India (1993): "Ground Water Resources and Development Prospects in Chennai District, Tamil Nadu", Central Ground Water Board.

[23] (1996): India Infrastructure Report: Report of the Expert Committee on Commercialisation of Infrastructure, Delhi: National Council of Applied Economic Research.

[24] (1997): National Institute of Urban Affairs.

[25] (2001): "Ground Water Quality in Chennai City", Ground water News Letter, Government of Tamil Nadu, India, Public Works Department, October-December.

[26] (1996c): "National Buildings Organisation: Hand Book of Housing Statistics", Part I, Ministry of Urban Affairs and Employment New Delhi.

[27] (2001): India Infrastructure Report, "Water Supply" New Delhi.

[28] (2002) Ministry of Water Resources: National Water Policy, New Delhi.

[29] Government of Tamil Nadu (2001-02): "Water supply resources of CMWSSB, Policy Note".

[30] (2002-03): "Water supply resources of CMWSSB", Policy Note.

[31] (2003-04): "Water supply resources of CMWSSB", Policy Note.

[32] (2004-05): "Water supply resources of CMWSSB", Policy Note.

[33] (2004): "Chennai Water-Checking Sea water Intrusion" Citizen Charter.

[34] India Development Report, (2002): Oxford University Press, New Delhi.

[35] Industrial Potential Survey Report, Chennai (2001-2002): Resource Analysis, Ministry of Small Scale Industries, Government of India.

[36] Paul Appasamy et. al. (1995): "Water Allocation and Management in the Chennai Metropolitan Area", Madras Institute of Development Studies (MIDS), Chennai.

[37] Manual, Corporation of Madras (Chennai), (1957): Madras City Water Supply Water Works Department, A.P. (M) Private LTD. Madras.

[38] Master Plan for CMWSSB -2011, (1991).

[39] Miloon, Kothari, (2003): "Privatising human rights – the impact of globalisation on adequate housing, water and sanitation" IMF and WB.

[40] Ninth Five-Year Plan, Government of India: "Drinking Water Supply and Sanitation of Tamil Nadu", India. [41] Planning Commission, India (2002): "India Assessment 2002: Water Supply and Sanitation".

[42] Review of Eighth Five-Year Plan (1992-97): Water Supply and Sanitation, New Delhi.

[43] Ruet J et Zû™ah M H (2001): "Water Supply and Sanitation in Indian Metros: Bombay, Calcutta, Madras", Research Report, New Delhi: Centre de Sciences Humaines.

[44] Sakthivel.R. (2003): "Metropolitan Water Use Conflicts: An Overview", Compendium Articles: Untied Nations Human Settlements Programme (UN-HABITAT) Information Office, Chennai City.

[45] Satterthwaite, D. and C.Tacoli. (2003): "The Urban Part of Rural Development: The Role of Small and Intermediate Urban Centres in Rural and Regional Development and Poverty Reduction", IIED.

[46] Sen, Indrani, (1990): "Country Issues paper: India: Financial Viability and Management' in Water Supply and

Sanitation", Beyond the Decade, Asia and Pacific Regional Consultation.4-8 June 1990, Asian Development Bank, Manila.

[47] Shobhnath, (2003): "Water Supply for Chennai City-Ground Water scenario". Compendium Articles: Untied Nations Human Settlements Programme (UN-HABITAT) Information office, Chennai City.

[48] Tamil Nadu Pollution Control Board (TNPCB), (2001): "Environmental Improvement of Water ways in Chennai City", Chennai.

[49] The Global Report, (2000): "Volumes 1, 2, and 3", Washington: U.S.

[50] UN report, (1997): "Ground water: Managing the "invisible" resource. United Nations", Educational, Scientific and Cultural Organization (UNESCO) Publication, Environment and Development Briefs, No. 2.

[51] UNDP-World Bank (1999): "Willing to Pay but Unwilling to Charge, Do "Willingness to Pay", Studies make a difference? Field Note, New Delhi: UNDP-World Bank Water and Sanitation Program, South Asia.

[52] UNEP (United Nations Environment Programme), (1982): "Environment and Development in Asia and the Pacific", Nairobi: UNEP.

[53] Vaidya Chetan, (1994): A study on willingness to pay for water and sanitation services: Case study of Baroda, Draft Report, Human Settlement Management Institute, New Delhi.

[54] Vande, Hock, Wim (2001): "Emerging water quality problems in developing countries", 2020 Focus 9, Brief 4 of 14, International Food Policy Research Institute, Washington.D.C.

[55] Virjee, K, (2003): The Willingness to pay for changes in Water, Wastewater and Electricity Services in Trinidad and Tobago, Draft Final Report to the Regulated Industries Commission of Trinidad and Tobago.

[56] Whittington, D., Briscoe, J and Mu., (1987): "Willingness to pay for water in rural areas: methodological approaches and an application in Haiti," Wash Field Report No. 213.

[57] WHO and UNICEF (World Health Organization and United Nations International Children's Emergency Fund) (2000): "The Global Water supply and Sanitation Assessment", New York.

[58] World Bank (1994): "World Development Report 1994: Infrastructure for Development", New York: Oxford University Press, London.

[59] (1996): "Urban Infrastructure Services Review, (Draft), Country Operations, Industry and Finance Division, Washington D.C: World Bank.

[60] World Development Report, (2004): "Drinking Water, Sanitation, and Electricity", London. [61] World Health Organization (WHO), (1993): "Guidelines for drinking-water quality: Volume 1 Recommendations 2nd edition", WHO, Geneva (2ndedition) Switzerland.

[62] WSP (1999): "Improving Water Services through Small Scale Private Providers: Water Vending in Chennai", Water and Sanitation Program, South Asia.

[63] World Health Report (2002) - Reducing Risks, Promoting Healthy Life.

[64] International Residential Code IRC, (2003): One and Two Family Dwellings, Publisher: ICC Product Type: 2003 International Residential Codes.

[65] Central Water Commission, (1993) "Water and Related Statistics, New Delhi: Statistical Directorate, Central Water Commission.

[66] India Infrastructure Report (2004) Recent Financial Initiatives Among Urban Local Bodies:

[67] A Report NIUA (2000) Norms and Standards of Municipal Basic Services in India

[68] India infrastructure Report (1991) Policy imperatives for growth and welfare, Published in 1996, Published on behalf

of the Ministry of Finance, Govt. of India by National Council of Applied Economic Research, New Delhi.

WORKING PAPERS

[1] Anand P. B. (2001): "Consumer Preferences for Water Supply? An Application of Choice Models to Urban India", Discussion Paper No. 2001/145 Copyright _ UNU/WIDER 2001, Center for International Development, University of Bradford, p.b.anand@bradford.ac.uk.

[2] (2001a): 'Water "Scarcity" in Chennai, India: Institutions, Entitlements and Aspects of Inequality in Access'. WIDER Discussion Paper No. 140. Helsinki: UNU/WIDER.

[3] Ayibotele, N.B. (1992): "The World's Water: Assessing the Resource". Paper presented at the international conference on Water and the Environment: Development Issue for the 21st century, January 26-31, Dublin Ireland.

[4] Bachrach, M. and W. J. Vaughan (1994): "Household Water Demand Estimation", Inter-Development American Bank. Washington D.C, Working Paper ENP106.

[5] David Hensher (2004): Households' Willingness to Pay for Water Service Attributes, University of Sydney,

[6] Dirgha Nidhi Tiwari, (2005): "Determining Economic Value of Irrigation Water: Comparison of willingness to Pay and Indirect Valuation Approaches as A Measure of sustainable Resource Use", CSERGE Working Paper GEC 98, Center for Social and Economic Research on the Global Environment University College London and University of East Anglia.

[7] Dublin Conference, (1992): "International conference on Water and Environment", 26-31 January, Dublin, Ireland.

[8] Foster, Stephen (1999): "The Interdependence of GROUNDWATER and urbanization", Proceedings of the Stockholm water symposium, Stockholm, August 9-12.

[9] Institute of Water Studies (1995): "River Basin Study on Chennai Basin Group an Abstract: Tharamani", Chennai, India.

[10] Janakarajan S and Marcus Moench, (2003): "Are Wells a Potential Threat to Farmers' Well being? The Case of Deteriorating Ground Water Irrigation in Tamil Nadu", Working paper no.174 of Madras Institute of Development Studies, Chennai.

[11] Lee,Kyu Sik and Anas, Alex. (1989): "Manufacturer's Responses to Infrastructure Deficiencies in Nigeria: Private Alternatives and Policy Options", Discussion Paper, Report INU 50, Infrastructure and Urban Development Department, World Bank, Washington, D.C.

[12] Mohanakrishnan. A. (1990): "Surface Water Availability in and around Chennai City", the Irrigation Management Training Institute, Tiruchy, Tamil Nadu, India.

[13] Pasumalaidevan. N. (2001): "Urban Water Supply-Chennai City-An Over View", Institute for Water Studies, Tharamani, Chennai.

[14] Sethi Kavita, (1992): "Households' responses to unreliable water supply in Jamshedpur, India: A Report", Draft working paper, Transport, water and urban development department, World Bank, Washington D.C.

[15] Sharma, Dev (2000): "Resource conflict-Water" Research Paper, CARE, Canada.

NEWS PAPERS

[1] The Hindu (2001): "Frontline, India's National Magazine from the publishers of The Hindu", Volume 18 -Issue 13, Jun. 23 - Jul. 06.

[2] The Hindu, 27th May (2002): "Demand for Canned Drinking Water Increasing".

[3] The Hindu, 7th July 2000.

[4] The Hindu, 7th July, (2004): Tackling Water Scarcity: "Government Floats Company".

Index

Z